PO… …CK

WIGTOWNSHIRE RAILWAYS

by

C.E.J. Fryer

THE OAKWOOD PRESS

ISBN 0 85361 408 3

Typeset by Gem Publishing Company, Brightwell, Wallingford, Oxfordshire.

Printed by Alpha Print, Witney, Oxfordshire.

Acknowledgements

I am indebted to the Mitchell Library of Glasgow, the Stranraer and Dumfries Public Libraries, the Scottish Record Office and the National Library of Scotland, Edinburgh, and the National Railway Museum Library, York, for opportunities for research and (in some cases) for the provision of illustrations. My special thanks go to Mr and Mrs H. Cheetham of Palnackie, for information supplied verbally in the course of a visit, and for permission to reproduce a number of superb photographs of the disused line.

Title page: A view of the beach at Portpatrick as seen on a Raphael Tuck commercial postcard *Oakwood Collection*

Bibliography

Boolls, Ethel M.: *Whithorn and the Southern Machars*
Cornwall, H.J. Campbell: *Forty Years of Caledonian Locomotives: 1882–1922*
Crofts, Freeman Wills: *Sir John Magill's Last Journey*
Highet, Campbell: *The Glasgow and South Western Railway*
Highet, Campbell: *Scottish Locomotive History: 1831–1923*
MacHaffie, G.: *The Short Sea Route*
Smith, D.L.: *The Little Railways of South West Scotland*
Smith, D.L.: *Locomotives of the Glasgow and South Western Railway*
Smith, D.L.: *Tales of the Glasgow and South Western Railway*
Thorne, H.D.: *Rails to Portpatrick*

Published by
The OAKWOOD PRESS
P.O.Box 122, Headington, Oxford.

Contents

Stranraer–Dumfries train crossing the Ken Viaduct shortly before the closure of the line, headed by a 'Black Five' 4–6–0. *E.R. Morten*

A busy scene at Newton Stewart station as captured in 1900. Note the Whithorn and Wigtown train is standing on the extreme right. *Courtesy Railway Magazine*

A 1900s view of Stranraer Town station. *Courtesy Railway Magazine*

Introduction

Galloway is not an area which receives much attention from visitors, although people who know it well revisit it again and again. Motorists from the south, touring Scotland for their annual holidays, usually give it a miss, leaving it well on their left as they travel northwards along the A74. The scenic attractions, they suppose, all lie north of the Central Belt. Certainly the hills over 3,000 feet high, the longest lochs, the most impressive waterfalls and the most celebrated fishing rivers are in the North – and also, one might add, the fiercest midges, Scotland's own brand of miniature mosquito, which ravages there during the summer months and extorts its tribute of blood. Galloway does not pretend to compare with Argyll, Breadalbane, Lochaber, the Cairngorms or the wild north-west, but it does have within a fairly small compass sights worth seeing and memories worth recalling, and in the buildings and ruins to be found there it exhibits a wider sweep of history than the more famous Highlands.

History came late to the latter, where for many a long year the royal writ from Edinburgh never ran. Up to the 16th century there are only legends, mostly bloody ones, and saints' lives of doubtful authenticity, to indicate what happened there. Clan chiefs within their own valleys exercised paternal authority over clansmen who bore their names and carried on feuds with chiefs of neighbouring territories. The distant government at Edinburgh scarcely touched their lives. They spoke Gaelic, not any Scottish version of the English tongue. Not till the time of the Massacre of Glencoe and the later Jacobite revolts did the policies of the South much affect them. The visitor will find few old buildings there, and such few towns as exist grew up in recent times.

It is otherwise in Galloway, which shares with the Border counties and the Southern Uplands a history of involvement in national affairs. Prehistoric settlers from the Mediterranean sailed north from the Straits of Gibraltar and left signs of their cultures in monuments of many kinds as they settled along the Galloway coastline. Stone circles, chambered tombs and monoliths can be found in large numbers. During the Iron Age brochs and hill-forts arose whose ruins are still to be seen. Roman armies came here, and in dry weather traces of their marching camps can still be detected in crop-markings. Later on Christian missionary monks left signs of their presence in standing crosses and inscribed tombstones. In the early 12th century the Norman invasion which began in England in 1066 reached Galloway and left its mark in many earthen 'mottes' shaped like enormous Christmas puddings which were once the centres of stockaded fortresses. Somewhat later the Cistercians and other religious orders established monasteries whose remains in two instances, Sweetheart and Dundrennan Abbeys, are as impressive as any ruins south of the Border. There are also many late mediaeval castles and fortified houses.

As for the inhabitants: during the Middle Ages the Scottish kings obtained some of their best soldiers from these parts. In the early 14th century Robert the Bruce won the first of his victories against the English in remote Glen Trool. During the later 17th century the uncompromising Calvinist Covenanters held their illegal outdoor conventicles here in lonely places, when people foregathered to listen to fierce pulpit oratory while some of their number kept watch for questing red-coated dragoons who prowled around looking for unlawful assemblies, meting out short shrift to offenders when they found them. Here and there one finds an inscribed stone saying that a godly man on this spot was foully murdered (or, from the government's point of view, a recalcitrant sectary was justifiably executed). In Wigtown Bay an old woman of 80 and a girl in her 'teens were tied to stakes in the sand at low tide and left to drown in the rising waters for supporting the cause of the Covenant.'

Galloway has literary associations – in particular with Robert Louis Stevenson whose two finest novels, *The Master of Ballantrae* and the unfinished *Weir of Hermiston*, are set in Galloway. Robert Burns, too, though he wrote mostly about neighbouring Ayrshire, at times strayed into this area. Some hymns in the Church of Scotland's hymnary are rhymed versions of extracts from the pulpit eloquence of Samuel Rutherford, once minister of Anwoth near Gatehouse of Fleet. John Buchan's book *The Thirty-nine Steps* has his hero Richard Hannay setting out for Newton Stewart from Euston, but alighting from the train at an earlier stop, 'a little place whose name I scarcely noted, set right in the heart of a bog' which may well have been the passing place at Loch Skerrow, but could perhaps have been Gatehouse of Fleet; from there he tramped across the Scottish lowlands with assassins at his heels, or surveying him from an aeroplane. Dorothy Sayers set a detective story, *The Five Red Herrings*, in and around Kirkcudbright, where Lord Peter Wimsey and his faithful manservant Mervyn Bunter tracked down the killer of a local and rather unpleasant artist, largely through discovering the railway journeys he made in order to establish an alibi. Finally one should mention Freeman Wills Crofts' readable detective story, *Sir John Magill's Last Journey*, in which Inspector French elucidates what happened to a business magnate who vanished from a sleeping compartment on the night express from Euston to Stranraer.

This book describes the construction and operation, more especially over the first 60 years when it had an independent or quasi-independent existence, of the railway that crossed Galloway from east to west, not only to link the area with the rest of the United Kingdom, but also to provide an additional connection with Ulster by way of the short sea route across the North Channel. It not only had stopping trains serving the area, but through trains also from London, with vestibuled coaches and sleeping cars, travelling overnight to connect with the cross-channel boats between Stranraer and Larne. Eventually its doom was decreed, and the only part of it now left is the short stretch between a point east of Dunragit and Stranraer Harbour, a mere six miles out of what had once been, branches included, a route mileage of 80. The rest has now reverted to the wild nature from which it was originally carved. One or two of its viaducts are still standing, and the line of the track-bed is for the most part still plain to see, though actually trying to walk along it is a difficult matter. The track, stations and other buildings, and the signalling apparatus have now largely vanished.

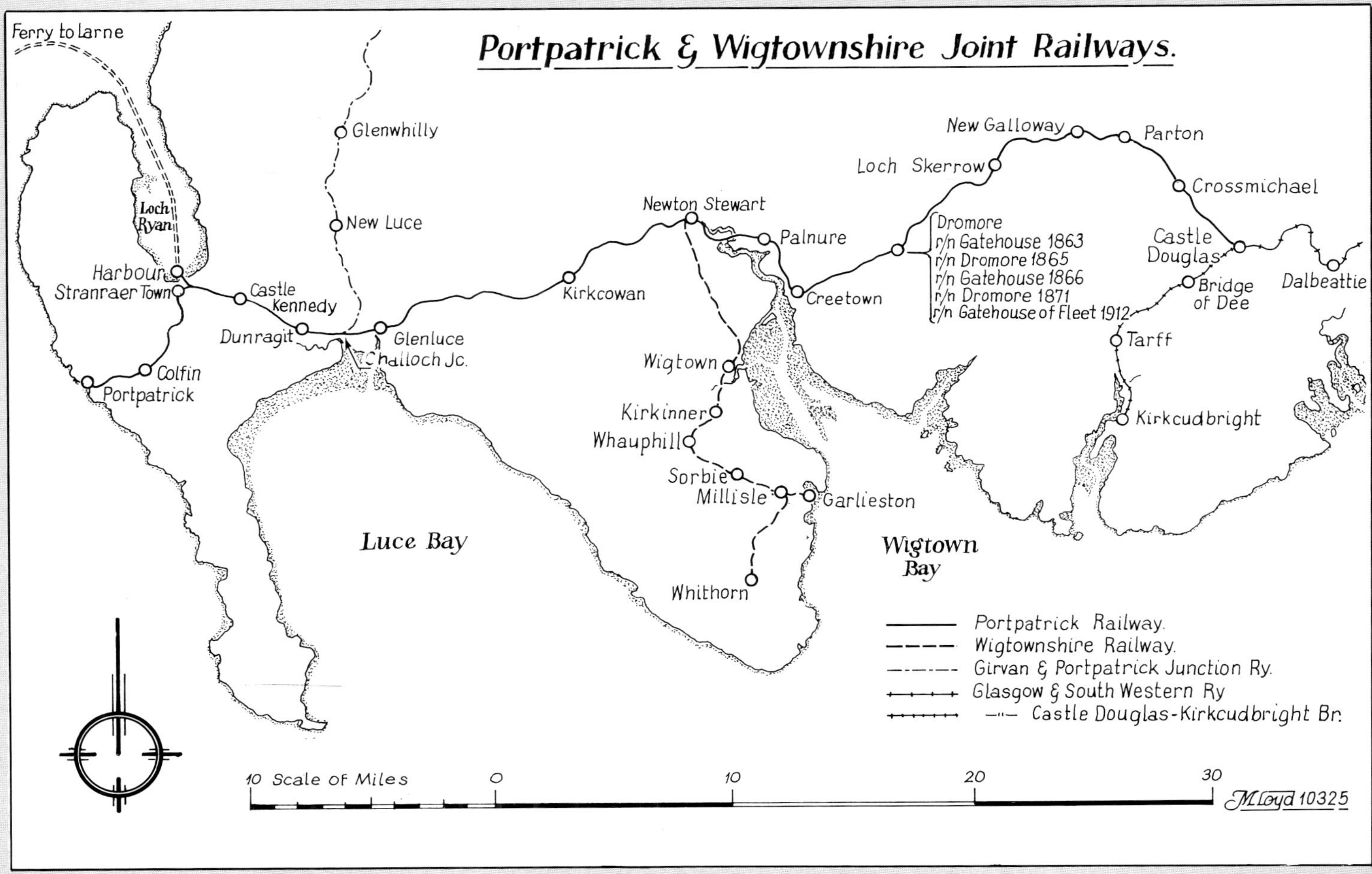
Portpatrick & Wigtownshire Joint Railways.
Ferry to Larne
Loch Ryan
Harbour
Stranraer Town
Castle Kennedy
Dunragit
Glenluce
Challoch Jc.
Colfin
Portpatrick
Glenwhilly
New Luce
Kirkcowan
Newton Stewart
Palnure
Creetown
Wigtown
Kirkinner
Whauphill
Sorbie
Millisle
Garlieston
Whithorn
Luce Bay
Wigtown Bay
Loch Skerrow
New Galloway
Parton
Crossmichael
Castle Douglas
Bridge of Dee
Dalbeattie
Tarff
Kirkcudbright
Dromore
r/n Gatehouse 1863
r/n Dromore 1865
r/n Gatehouse 1866
r/n Dromore 1871
r/n Gatehouse of Fleet 1912
Portpatrick Railway.
Wigtownshire Railway.
Girvan & Portpatrick Junction Ry.
Glasgow & South Western Ry
—"— Castle Douglas-Kirkcudbright Br.
10 Scale of Miles
0
10
20
30
M.Lloyd 10325

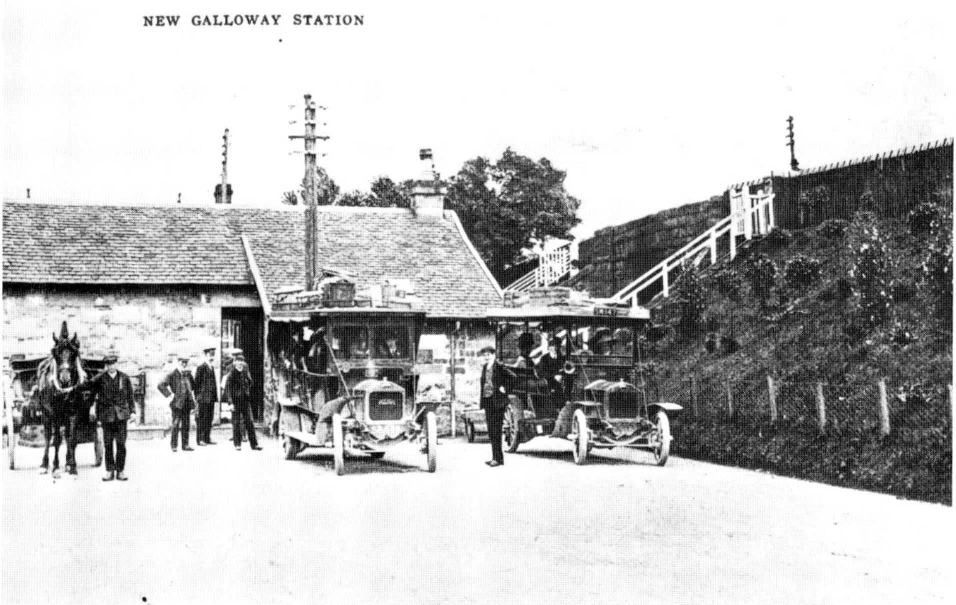

New Galloway station courtyard seen here in 1900. The two motor wagonettes and horse and trap are waiting to convey the passengers to New Galloway town. Note the solid tyres on the vehicles and 'hooter' style horns. *Lens of Sutton*

A view overlooking Gatehouse towards the Fleet Estuary and Wigtown Bay. *Oakwood Collection*

Chapter One

The Region Served

The Portpatrick and Wigtownshire Railways served, in a restricted sense, the shoreward areas of Kirkcudbrightshire and Wigtownshire, and in a wider sense the whole United Kingdom, since it provided a short sea route to Northern Ireland. It, together with the branch from Dumfries to Castle Douglas which gave access to it, was the only railway within these two counties until 1877, when the Girvan and Portpatrick Junction line for the first time linked Glasgow and Ayrshire directly with Stranraer. This latter railway is now the only one left to pass through Wigtownshire, while the Stewartry of Kirkcudbright (pronounced Kakoobrey) now shares with the shires of Peebles, Selkirk and Roxburgh the doubtful distinction of including no railway at all.

The boundary of these two adjacent county areas, which have now lost their separate existences and been merged in a larger unit, but which were administrative realities at the time with which this book deals, leaves the sea coast at Loch Ryan and strikes eastward across largely uninhabited moorland until it reaches the river Cree. This, one of the principal streams in this region, takes a more or less south-eastward course to its estuary at the head of Wigtown Bay. The dividing line between the two counties follows this river to its mouth, while the border between the Stewartry and the shires to the north goes along the Cree upstream to its source on Kirriereoch Hill. Here it turns sharply southwards to Loch Enoch in the wildest part of the Galloway hills, almost doubles back on itself by following the river Doon as far as the loch of that name, from which it turns away midway along the latter's eastern shore, going north-eastwards and then eastwards to the summit of Alhang, at 2,100 ft the highest point along the course. It then continues south by east to a point west of the burgh of Dumfries, cuts across eastwards almost to the river Nith, curves to the south to avoid Dumfries itself and so reaches the Nith estuary. For the rest of its course it follows the coast of the Irish Sea and North Channel, around the Urr, Dee, Fleet and Cree estuaries and the southward-projecting peninsula of the Wigtownshire Machars to the mouth of the Water of Luce; it then follows the shore again to the southernmost point of the Rhinns of Galloway, where a prominent lighthouse crowns an eminence at the end of a line of savage cliffs, and then continues up the west shore of the Rhinns past Portpatrick and round Milleur Point to Loch Ryan again.

This sinuous boundary encloses much uncultivated moorland, in places as wild as any the Northern Highlands can show, and almost as bleak in winter, with hill summits as fine as those in the English Lake District (though far less often visited), many fresh water lochs, many rivers filled with fish and once more the habitats of recently-reintroduced and protected otters, and one famous beauty spot, Glen Trool. In the more southern parts, around the coastline and inland along the river valleys, there is good farming land, broken here and there with patches of heath, and a sprinkling of villages and small townships, well connected by metalled roads. Of late there has been much afforestation on the fringes of the higher ground, but when the railways were first constructed this had not begun.

Nearly all the towns are in the river valleys, at the heads of estuaries or on the sea coast, Whithorn being the only sizeable exception. The shores have many fine sandy beaches and rocky outcrops appear almost everywhere, rising in places into fine cliffs, notably westwards and north-westwards for some miles from Burrow Head at the southern extremity of the Machars, and in the same direction from the Mull of Galloway. The latter is one of the most impressive headlands in the United Kingdom; on a clear day the views from its lighthouse are magnificent, extending to the Irish coast and the Isle of Man. At the Mull one is at the southernmost point of Scotland, in the same latitude as Hartlepool on the English east coast. The Rhinns of Galloway have a character all their own; the hills are low, green and much broken, with whitewashed farms nestling in the hollows. They almost constitute an island, being linked to the mainland by a flat sandy neck only six miles across, on the northern edge of which, at the south end of Loch Ryan, is Stranraer, the principal town of the region and the port for Larne in Ulster. Vehicle-carrying ferry-boats now ply the 40-or-so miles between the two towns several times each day.

The main lines across the English–Scottish Border had already been built when the demand began to be heard for a rail link between Dumfries and Galloway. In the early 19th century road transport was necessarily slow. Passengers were carried in vehicles drawn by horses which had to be changed every so many miles, and travelling speeds averaged about seven or eight miles an hour. Dumfries was a long day's journey from Stranraer. A few small ships crossed the Irish Sea and the Solway Firth to ports such as Whitehaven and Liverpool, but these were as slow or slower than the road coaches, though more convenient for the transport of heavy goods. In regard to time expended getting there, Galloway was as far from England as Aberdeen or Inverness. It had, and still has no heavy industries, but was a potential source of farm and dairy products which, given better communication, might have been expected to find markets in the larger urban centres of southern Scotland and northern England.

Galloway was also a land bridge stretching almost the whole way from the Anglo-Scottish border to Ulster. The straits between Portpatrick and Donaghadee in County Down are no wider than those between Dover and Calais. So any railway, once it had been built, was bound to have far more than a local importance. Since 1662 the Post Office mails had been shipped between Portpatrick and Donaghadee, and although the British Government ended this service in 1849 it was soon to re-commence, though not, as things turned out, from Portpatrick. Once the Portpatrick Railway had been built it served Northern Ireland as well as Galloway, and was valued accordingly. The 'Short Sea Route' offered definite advantages to the traveller – even to one beginning his journey in the south. Once the through train service had been introduced, first on the Midland and then on the West Coast route, it took no longer to travel overnight from London to Belfast by way of Stranraer and Larne than to go by the earlier-established route through Holyhead and Kingstown, if one were proceeding to Belfast or its vicinity, and even before the introduction of sleeping cars it was probably less uncomfortable.

A passenger from London bound for Northern Ireland had a less troublesome ride if he boarded a train for Carlisle in the early evening, and merely crossed a platform there during the small hours to make the Stranraer connection, with the prospect of having breakfast on board the steamer and freshening himself up with razor, soap and toothbrush before disembarking at Larne for the final short stage to the Ulster capital. Once sleeping carriages had been introduced on the Stranraer run there was no question about it, so far as the first-class traveller was concerned. Even though he might not sleep soundly, he could at least doze in a recumbent position. As for passengers from Scotland to Ulster, they had the alternative of direct boats from the Clyde to Belfast, but after 1877 the route by Ayr and Girvan was open, and during the present century it has been possible to take the late night 'Paddy' from St Enoch, Glasgow, book a berth on the morning boat and sleep on board while it was still in harbour. (This, alas, is no longer possible since the boat does not reach Stranraer from Larne in time.) So for the Scots, too, the short sea route was extremely convenient.

A view of the fishing fleet in Portpatrick Harbour. *Oakwood Collection*

Chapter Two

The Castle Douglas and Dumfries Railway

Though not itself a part of the Portpatrick Railway which was later to be built, the line from Dumfries to Castle Douglas was the necessary link that enabled through traffic to pass from other parts of the United Kingdom to Galloway. During the middle 1850s demands for the building of such a line began to make themselves heard.

The Glasgow, Dumfries and Carlisle line, later to become the principal main line of the Glasgow and South Western Railway, had been opened in 1850. Between it and the Caledonian Railway, which competed with it for the traffic between Glasgow and the south, a strong rivalry developed which continued after its absorption into the G&SWR and motivated the latter to consider making a line from Dumfries to the south-west in order to forestall the CR. However, no steps were immediately taken, since it was plain that the latter had enough on its hands elsewhere to engage its energies at that time, and it was not until 1853 that the project was again mooted.

The inducements were both the profits to be expected from bringing railway services to hitherto unreached areas, and the prospect of advantages to be gained from tapping traffic to and from Northern Ireland – in particular the mail traffic. Mails from the northern parts of the United Kingdom to Ulster had for a long while been taken by road coaches to Portpatrick, shipped thence to Donaghadee and then carried on by road to Belfast. It was a long and slow journey by road, though a short one by water. At the beginning of the 19th century Portpatrick was a busy place, with mail packet vessels crossing regularly to Ireland, to which place passengers and cattle were also conveyed. As many as 10,000 letters a day were taken across in both directions. It was also the recognised route for despatching troops to Ireland, and barracks had been built at Portpatrick to accommodate soldiers prior to embarkation. Under pressure from the Government the harbour had been improved. However, during the 1840s there was a change of policy. The opening of the railway from Crewe to Holyhead made possible a direct rail-and-steamer linkage with Dublin, and in September 1849 mail traffic was withdrawn from the Portpatrick–Donaghadee route since it had no rail links on either side of the North Channel. Could such links be provided, it was made known, the Government would think again. However, in that year the nearest rail-heads to Portpatrick were Ayr and Dumfries, 60 and 80 miles away respectively.

In their report to their shareholders in March 1853 the Directors of the G&SWR (which had now absorbed the Glasgow, Dumfries and Carlisle line) commended two proposals, one for a line between Ayr and Girvan and another between Castle Douglas and Dumfries. They were planning a 'pincer movement' for the invasion of Galloway by rail, and these were the first stages, the intention being eventually to reach Portpatrick and provide the Post Office with the rail links it needed from Scotland and the North of England to Ulster. A third proposal was also put forward, for establishing a steamer service, owned and organised by the railway, from Ardrossan to Belfast. This would merely have involved making a short branch from the existing line linking Glasgow with Ayr, and the shareholders approved the suggestion, but Parliament turned it down.

Meanwhile the Castle Douglas and Dumfries scheme was being promoted locally. Funds, however, proved difficult to raise, so the G&SWR took a more active interest in it and offered to contribute half the cost of constructing the line, or £60,000 whichever was the less. The promoters approved and a Bill was put before Parliament; it attracted hardly any opposition and was passed in July 1856. Twelve local landowners were named in the Bill as promoters, and the authorised capital was £120,000, issued in £10 shares. The first shareholders' meeting was held in the Court House at Castle Douglas on 5th September, 1856, and the Directors' first report was read. Portions are quoted below:

> The want of railway communication in this part of Scotland has been one long and deeply felt . . . Some efforts were made at a previous period to open up the district, but nothing was actually accomplished till 1846, when an Act was passed for making a railway in connection with the then Glasgow, Dumfries and Carlisle Railway, from Dumfries to Kirkcudbright, with a branch to Gibbshole at the mouth of the Water of Urr; which, however, was abandoned in consequence of the financial difficulties which so soon followed upon that eventful period.
>
> No further steps were taken . . . till about three years ago, when a survey was made of a line of railway from Dumfries *via* Dalbeattie to Castle Douglas, but the project was . . . for the time postponed and nothing further was done till last autumn when, it being understood that the Glasgow and South Western Railway Company . . . would aid liberally any undertaking which might receive general support from proprietors and others locally interested, the project was again set on foot . . . It was resolved to make every effort to raise the means necessary for applying to Parliament . . . and the Bill for the Railway was duly presented . . .
>
> The length of the line is 19 miles 3 furlongs and 77 yards, and the capital is £120,000; one half of this was subscribed by the general public, and the other half by the Glasgow and South Western Railway Company. The Company was thus placed in the position of having their whole capital subscribed, a position which few companies have attained at so early a stage of their proceedings . . .
>
> Negotiations were entered into with the proprietors of the land on the line, and the Directors have already secured nearly three-fourths of the land . . . in most cases at its agricultural value . . . The Directors are satisfied that they have succeeded in economising the funds of the Company . . . This Act received the Royal Assent in July last.
>
> The Directors have much pleasure in adverting to the projected railway announced by the landed proprietors and others in Wigtownshire . . . The Government recognise the great advantages which would be derived by establishing a communication between the ports of Portpatrick and Donaghadee, thereby reducing the sea passage to the shortest possible time, and have agreed to adopt that route as a mail passage . . . on the condition that the railways on each side of the Channel shall be completed, so as to connect Donaghadee on the one side with Belfast, and Portpatrick on the other side with Glasgow to the north and Dumfries to the east . . . The Directors hope that at no distant day an undertaking fraught with so many advantages will be realised.

A year later they were able to report that the first section of the line, between Dumfries and Lochanhead, had been let for £63,000 to Henry Lee, Son and Freeman, the length being just over 5½ miles. The same firm had also undertaken to continue the line to Dalbeattie, 7½ miles, for £25,000; the remainder to Castle Douglas, rather less than 6½ miles, was allotted to Layton Brothers of Newcastle-upon-Tyne, and this was to cost a little under £35,000.

However, once work had begun the expenditure went beyond the Company's means, and Parliament was petitioned for permission to raise new stock and borrow an additional £24,000. This was allowed, and the G&SWR was authorised to acquire all the new stock. The line was duly completed and opened for traffic on 7th November, 1860. The following September a dividend of 5 per cent was declared on preference stock and 1 per cent on the original shares. Traffic continued to increase, and was greatly augmented when the projected Portpatrick Railway was opened at the end of the same year. As receipts mounted, dividends increased. The G&SWR had from the start the largest interest in the line, and in 1865 it took it over.

As the costs quoted above show, the first stage from Dumfries to Lochanhead was the most expensive length to construct. An inspection of the Ordnance Survey Map of the area shows why. Heavy earthworks were required through difficult terrain, and a rocky, steep-sided pass had to be negotiated as Lochanhead was approached; the wide river Nith had also to be bridged soon after the junction with the main line north of Dumfries. From Lochanhead onwards the route follows the south side of a wide valley, aiming south-west, and not many excavations were required until Dalbeattie was reached. At the latter town the line swung round to the north in a wide semicircle around a ridge of hills, followed the Urr Water, crossed it and then turned west towards Castle Douglas.

Meanwhile the aspirations of the chief town in the Stewartry were not forgotten. Kirkcudbright had been the line's original goal, and money came in satisfactorily enough to enable the continuation from Castle Douglas to be completed, though the branch to Gibbshole never materialised, and was perhaps never really intended but had been included in the original Bill to put a brake on the ambitions of the Caledonian. Work began in the summer of 1862 and was completed in March 1864. This continuation cost £60,000, two-thirds of which was raised locally, the G&SWR and the Castle Douglas and Dumfries Railway contributing the remainder. It was just over 10 miles long, and crossed the Dee twice by large bridges, the building of which used up a large part of the available money. A year after the project was complete the G&SWR took the line over, together with the CD&D.

The 3-span Ken Bridge, near Parton (*details page 136*). *Courtesy Railway Magazine*

Chapter Three

Building the line to Portpatrick

The moving spirits behind the construction of the Portpatrick Railway were two local dignitaries: Lord Dalrymple, Lord Lieutenant of Wigtownshire, and Sir William Dunbar, seventh Baronet of Mochrum, M.P. for the Wigtown Burghs from 1857 to 1880 and a keen supporter of railway development in all parts of Scotland.

At a County meeting in Wigtown on 30th April, 1856 a letter from Lord Dalrymple was read in which he suggested that enquiries be made to discover what support there might be for a railway through Galloway, which would not only be of local benefit but would also complete the link between existing lines in Great Britain and the coast facing Ireland at its nearest point. It would connect Galloway with the rest of the Scottish railway system by making a junction with the Castle Douglas and Dumfries Railway (already mooted though not yet built). A Committee was formed to look into the matter, and further to enquire whether a particular engineering firm, E. & E. Blyth of Edinburgh, would be prepared to survey and, if so desired, construct such a line. Meanwhile arrangements were made to raise money.

Six weeks later, at a public meeting at Newton Stewart on 13th June, the surveying of the proposed line was approved. On 19th September, again at Newton Stewart, Mr E. Blyth reported his surveyors' findings. It was agreed after some discussion that the line should pass through Stranraer and not just be linked to it by a branch, since a number of railways which were interested in the project and wished it success had given a strong indication that they expected the line to serve that town directly. It was also stated that many landowners in the region were prepared to sell the necessary land at its agricultural value. This would mean a significant economy, so it was also proposed and agreed to buy enough land to allow for a double line of track, though at first it was to be single. Local committees were appointed to persuade people to purchase £10 shares. There was some urgency in regard to this, as the railway companies interested in the project were insisting that at least £240,000 must be raised locally by the end of the year if they were themselves to contribute. In fact this objective was not quite achieved, and it began to look as if the other promised funds would not be forthcoming.

Then something unexpected happened; the Great Northern Railway of England offered to invest £160,000. This needs some explaining. What interest could a company in the south of England have in a railway in Galloway? But the GNR was playing for high stakes. It had already reached Leeds and Bradford. Beyond those cities a smaller company friendly to the GNR – later to be absorbed into the Midland Railway but as yet independent – the so-called 'Little North Western', had reached Lancaster from Leeds and was building a branch from Clapham to follow the Lune valley and link up with the Lancaster and Carlisle Railway (later a part of the London and North Western system) near Tebay in Westmorland. So there was a possibility that traffic from London to Portpatrick, bound for Ulster, would pass first over the GNR and then over a line friendly to it to within 40 miles of Carlisle. The GNR was in fact making its contribution conditional on this route being followed from the south to Galloway.

The other lines which were interested in tapping the traffic to or from Galloway now became alarmed, and a good deal of discussion and bargaining went on after it was known that the new railway's promoters were inclined to consider the GNR's proposal favourably. In the end they secured the rejection of that offer, but it had had its usefulness in making the other companies more ready to contribute to the Portpatrick Railway's construction. Eventually a Bill was put forward, passed through Parliament and duly received the Royal Assent. Money had now been promised sufficient to complete the line, the Lancaster and Carlisle and the G&SWR undertaking between them to make up the amount which the GNR would not now be contributing.

According to the Act of Parliament, passed in 1857, the new line was to begin at a junction with the Castle Douglas and Dumfries Railway and to end at a point on the hillside above Portpatrick Harbour. From the latter a spur line was to descend to the harbour itself, while at Stranraer another branch would link that station with a pier on Loch Ryan. The complete railway was to be built within 5 years. The Treasury also undertook to pay for the necessary improvements at Portpatrick Harbour and at Donaghadee on the Irish coast to enable the packet service to be instituted.

So far, so good – but now dissension arose over the route the line should take. Between Castle Douglas and Creetown there were two routes possible, one aiming at Kirkcudbright (which had not yet been reached by the branch later built from Castle Douglas) and then crossing the Dee estuary, subsequently following the coast through Gatehouse of Fleet and so continuing to Newton Stewart. The northern route, shorter though more hilly, would pass up the valley of the Dee to the ferry-point at Boat of Rhone, cross Loch Ken by a bridge at its narrowest point, and then strike across uninhabited moorland towards Creetown, leaving Kirkcudbright and Gatehouse of Fleet well to the south. The surveyors favoured the northern route, but there was a strong contrary feeling expressed at the Company's second general meeting in February 1858 that Kirkcudbright and Gatehouse should not be passed by. However, the surveyors' report insisted that the southern route would be harder to construct, with more earthworks, rock cuttings, bridges and a 500-yard tunnel. It would cost £70,000 more, and any additional traffic gained would not compensate for the extra expense. Sir William Dunbar turned the opinion of the meeting when he reminded it that the line was primarily to be a link with Ireland, and that towns left on one side could if they wished promote rail links of their own to join it.

Tenders were sought for the eight separate contracts by which the line was to be built. These were duly submitted and in each case the lowest estimate was accepted. All were set in hand without delay except for the Stranraer–Portpatrick contract, which was postponed because the harbour was not yet ready to receive boats for the packet service to Donaghadee. However it had always been intended that heavy goods from or to Ulster should be landed or shipped at Stranraer, so that when the line had reached the latter port goods traffic could begin, and passengers might also use the facilities and travel to Belfast that way.

The contracts were awarded as follows:

Section	*Length*	*Contractor*	*Cost*
Castle Douglas to Boat of Rhone	7¼ m	Thos Telford, Mitchell & Co.	£26,000
Loch Ken Viaduct	½ m	Thos Nelson & Co.	£12,289
Loch Ken to Dromore	11½ m	McNaughton & Waddell	£95,840
Dromore to Palnure	7¾ m	James Cowans & Co.	£39,403
Palnure to Kirkcowan	9½ m	McDonald & Grieve	£46,895
Kirkcowan to Glenluce	8½ m	James Falshaw & Co.	£41,788
Glenluce to Stranraer	9¼ m	James Falshaw & Co.	£28,974

The high cost of the Loch Ken to Dromore contract was due to the intractable nature of the terrain and the need to build five large bridges, the two largest being many-arched viaducts over the Little and Big Waters of Fleet. The final stage, on the other hand, going through very easy country was correspondingly cheap to construct.

Work began on all seven sections in the spring of 1858. The weather that year was phenomenally fine until the end of the summer and construction went rapidly ahead. There was, however, an unfortunate delay at Castle Douglas. Here the Portpatrick Railway had meant to build its own station adjacent to that of the Castle Douglas and Dumfries line, but this would have made a road diversion necessary, turning a straight approach to the town into a crooked one, and the Stewartry authorities objected. It was late autumn before work in Castle Douglas could start, after the argument had been settled through a change of plan by the Portpatrick line, which agreed to share the other company's station, a much more sensible arrangement when through traffic was contemplated.

On many parts of the route work went on day and night, the heavy excavation being performed by 'navvies', many of them immigrant Irishmen attracted by the pay and willing to endure the primitive conditions. There was less rioting than had been expected. Policemen and special constables kept an eye on the small armies of diggers, and one Kirk minister in Glenluce thoughtfully organised a savings bank to encourage them not to spend their pay on whisky.

After the fine summer of 1859 the weather turned bad – so unpleasant indeed that many navvies deserted and went to seek work in less distressing conditions. The severe winter was followed by a wet summer. The tempo of construction slowed accordingly, but by the onset of autumn the rails were being laid and the end was in sight. In September 1860 the line received its first locomotive, the contractors on the Newton Stewart section taking delivery of a 12 ton tank engine which was needed for track-ballasting operations. It came by rail from Glasgow to Castle Douglas and was then partially dismantled and taken in four sections along the road to Newton Stewart. Its arrival there provoked much excitement, the last stage up to the station being allegedly accomplished 'at full gallop'. Once re-assembled at Newton Stewart it not only performed its allotted duties but also ran some excursion trains to Creetown, making five separate return trips.

By November 1860 the only breaks in the newly-laid line were on the difficult moorland section between Loch Ken and Dromore, and these were closed before the end of the year. In February 1861 the Board of Trade inspection took place, the Inspector, Captain Tyler, spending three days carefully examining the line and its structures and then declaring them satisfactory. Government permission for the opening followed, and this took place on Monday 11th March, 1861.

It was an occasion for much enthusiasm and the usual ceremonial junketing. The company had purchased two locomotives, inside-cylindered 0–4–2 mixed traffic engines, numbered 1 and 2. These hauled a special train carrying Directors and shareholders from Stranraer to Castle Douglas, leaving at 9 am and arriving at 11.30. On arrival cheering crowds greeted it and a salute of guns was fired. The Directors were then entertained to a luncheon at the Douglas Arms. Hunger being appeased, they and the other passengers returned to the train, which set off once more, to halt at Newton Stewart an hour and a half later; it then waited there for 3 hours while the travellers consumed yet another meal, a dinner in the Assembly Rooms. Speeches were made and toasts were drunk to the success of the new Railway, and there was much applause. At 5 pm the train re-started and was back at Stranraer before 6. The next day the line was opened for public service.

It was not until the summer of 1862 that the final stretch to Portpatrick was ready for opening, though work on it had begun three years earlier. The delay was partly due to constructional difficulties, partly to misunderstandings with the Government which, having undertaken to make Portpatrick and Donaghadee harbours suitable for the use of packet steamers, had been dragging its feet. In September 1859 the Portpatrick Railway Company took the matter up with the Board of Admiralty, insisting that before this section of the line could be built there had to be some definite understanding between the Company and the Government. There was no point in building a line, whose chief purpose was to convey traffic to and from Ireland, towards a port which could not receive the ships with which the trains were supposed to connect. The Admiralty replied that the necessary harbour works were about to commence. Six months later there was still no sign of this happening, so a further remonstrance was sent. Again a reassuring reply was received; a contract for the work had now been signed. However it was not until March 1861, when the line from Castle Douglas to Stranraer was already in operation, that the contractor, a certain 'Mr Greenhill from London', arrived at Portpatrick to commence operations there. By that time, relying on Government promises, the Portpatrick Company had begun the construction of the line.

The 7-mile stretch south-westwards from Stranraer was begun early in 1860, the contractor being Thomas Nelson who had also built the Loch Ken viaduct. His quotation for the work was £55,886. At this particular moment the Portpatrick Company was being financially squeezed by the G&SWR, which had contributed £60,000 towards the construction of the whole line and had promised a further £40,000. Thinking it had the smaller company in a cleft stick, the larger one tried to insist that the Belfast and County Down line, which was to provide the necessary final link from Donaghadee to Belfast, and which had already contributed £15,000 and promised a further

£20,000, should pay over the latter sum before it would part with its own £40,000, although it was already obvious that the Irish company was in no position to meet this obligation. The G&SWR supposed that the Portpatrick Company needed its co-operation to work the new line and would rely on it to supply locomotives and rolling stock. It was sadly mistaken. The Portpatrick Railway's reaction to this piece of attempted tyranny was to decide to do without the G&SWR's money, to provide its own engines and carriages and make ends meet somehow or other. This made it all the more necessary that the coveted Irish postal traffic would really come their way.

The length from Stranraer to Portpatrick was not easy to build. A long viaduct was needed to bridge the Piltanton burn a quarter of the way along the route. It was substantially built of local whinstone, with a maximum height of 73 ft above the burn. Cuttings on either side had to be driven through boulder clay, which proved no easy task. The final approach to the cliff-edge at Portpatrick involved much blasting away of rock before the site of the terminal station could be reached. The gradient profile shows the line to have been a steep gable, four miles up and three down, at 1 in 72 or steeper, with a summit at Colfin 326 ft above the sea. The advantage experienced when the other seven sections were built, of a fine spring and summer, were not enjoyed when this final stretch was constructed; the winter was cold and the summer wet. Unfortunate accidents accompanied the making of the viaduct. During July 1861 two of the arches crumbled and had to be reconstructed. Three months later, as a cart full of building stone was being taken across, part of the wooden staging collapsed and three young workers fell 60 feet to the ground below; two were killed outright and the third was badly injured. Rather later, when the line was almost finished, a crossing gate was being fixed at the summit at Colfin, which necessitated some blasting of rock; the charge exploded prematurely and two workmen were badly hurt; one lost an eye and the other had both legs broken.

However, on 1st August, 1861 the Board of Trade inspection took place and the line was passed for public service, which began four weeks later. It had, however, to be a purely local service, since the harbour works were not yet complete, though the spur line from Portpatrick station to the harbour had been finished. The dilatory Mr Greenhill had been taking his time, and there had been long delays in excavating the basin where the packet steamers were to berth. In June 1862 the Portpatrick Company complained about the delay. They had built their line to the harbour but, so far as the Irish traffic was concerned, to no effect. Fears were arising – all too well founded, as it turned out – that the port never would be adapted for the packet service.

It is perhaps worth considering the position from the Government's side. The Board of Admiralty, whose main concern was with defence, was not really interested in promoting the affairs of the Post Office. The latter, too, one may suppose, did not think that routing mails across the North Channel was a matter of vital importance. Steamers between Holyhead and Kingstown were now carrying mails from London and all parts of England and Wales to Dublin more rapidly than ever before, and Belfast was only 115 miles further along the coast. Only Scotland and parts of the north of

England would derive much benefit from a packet service on the short sea route. So were the harbour works at Portpatrick and Donaghadee really worth the expense – particularly when a sea link between Stranraer and Larne, admittedly longer than the Donaghadee route but not all that much longer, already existed and could be improved?

Assurances continued to be given, and feet went on being dragged. Meanwhile, now that Stranraer had been reached, the building of a short branch to the harbour was undertaken – not so much for the benefit of passengers to Ireland – they, it was still hoped, would go by way of Portpatrick – but for heavy goods traffic. The Stranraer Burgh Council constructed a pier 800 ft long, largely of timber, at which vessels of up to 12 ft draught could berth at all states of the tide, the Portpatrick Railway company agreeing to pay £1,100 annually for using it. Both pier and branch line were completed in time for the Portpatrick section and themselves to be inspected together, and the opening ceremony for the former, though a more subdued affair than the previous one, none the less occasioned local rejoicing, and a large party of dignitaries from Ireland, headed by Lord Dufferin, joined in the celebrations.

In 1865 Parliament granted a further £10,000 for the completion of Portpatrick Harbour, and matters livened up a little in consequence. At about this time the Board of Trade took over from the Admiralty the oversight of British harbours, and in 1866 it sent an engineer to Portpatrick to inspect the nearly-finished works. His report was unfavourable. The entrance, he said, was silting up, and was in any case too narrow for ships to negotiate in dark and stormy weather. The Board of Trade now had a revulsion of feeling against the whole project, and in November 1866 they remonstrated with the Treasury, asking to be relieved from 'wasting public money in carrying out an undertaking that was originally vicious and can never . . . be of any public service'.

Whether or not this was an over-reaction, the Portpatrick Railway Directors were unwilling to accept the situation. There was much local feeling in favour of the short sea passage, and even though the Government was prepared to recompense the Company for its disappointed expectations they resolved to have an independent inspection. Sir Luke Smithett of Dover, who had had 50 years of experience in packet service between Kent and the Continent, was asked to give his opinion. The report he made was warmly in favour of the port being used as originally intended. But the Government still regarded Portpatrick Harbour as a millstone round its neck. Instead of paying attention to Smithett's report it began to negotiate for a quicker rail passage by way of Dublin and Belfast. So the ridiculous situation was continued, of a letter from south-west Scotland to Belfast having to travel some 540 miles instead of a mere 60 or 70.

After a long negotiation between the Directors and the Government, in which the former did not fail to point out that their line had been primarily built to provide a link with Ulster, and that the section west of Stranraer would never have been built at all had it not been for official assurances that there *would* be a packet service from Portpatrick to Donaghadee, terms of compensation were agreed. The Government offered to pay the Portpatrick Railway £20,000 outright and make it a loan of £153,000 at 3½ per cent interest to cover the extent of its indebtedness. Nothing better could be extracted from Lord Derby and his Cabinet. So the Directors reluctantly accepted the suggested arrangement and Portpatrick was left to remain a pleasant and sleepy backwater. As will appear, its place was very successfully taken by Stranraer. There was never any fear that Loch Ryan would silt up.

CASTLE-DOUGLAS ETC., TO STRANRAER & PORT-PATRICK.

FARES from Castle-Douglas. 1 Cl. s. d.	2 Cl. s. d.	3 Cl. s. d.	Mls	Trains leave	a. m.	a. m.	a. m.	p. m.	
..	..	..	..	Carlisle v. Cal. ......at	..	6 15	8 20	1 58	..
..	..	..	..	Do. v. G.&S.W.	..	6 36	11 0	3 30	..
..	..	..	..	Edinburgh v. Cal.	..	..	10 0	1 0	..
..	..	..	..	Glasgow "	..	..	9 45	1 0	..
..	..	..	..	Do. v. G.&S.W.	..	..	8 55	10 30	..
..	..	..	..	Lockerbie	..	7 15	12 20	4 0	..
..	..	..	..	Dumfries	..	8 0	1 10	4 45	..
..	..	..	..	Classes........	1 2 3	1 2 3	1 2 3	1 2 3	..
..	..	..	..	Trains leave......	a m	a m	p m	p m	..
..	..	..	..	Cstle-Douglas ......at	..	9 0	2 5	5 45	..
0 7	0 6	0 3½	3½	Crossmichael	..	9 7	2 15	5 53	..
1 2	1 0	0 6½	6½	Parton	..	9 15	2 24	6 2	..
1 6	1 1	0 8½	8½	New Galloway	..	9 24	2 31	6 9	..
2 0	1 6	0 7	18½	Gatehouse Station †....	..	9 50	2 56	6 32	..
4 0	3 0	1 11	23½	Creetown	..	10 0	3 9	6 43	..
4 5	3 4	2 2½	26½	Palnure	..	10 8	3 19	6 52	..
5 0	3 9	2 5½	29½	Newton-Stewart †......	..	10 16	3 30	7 5	..
6 0	4 6	3 0	36	Kirkcowan	..	10 30	3 41	7 24	..
7 5	5 7	3 8½	44½	Glenluce	..	10 47	4 9	7 45	..
8 0	6 0	3 11½	47½	Dunragit	..	11 0	4 15	7 51	..
8 6	6 4	4 2½	50½	Castle-Kennedy	..	11 8	4 21	8 0	..
0	6 8	4 5	53½	Stranraer Station arr	..	11 15	4 30	8 10	..
				Stranraer Station lea	9 45	11 30	4 45	Stop	..
9 7	7 3	4 9½	57½	Colfin, ...	9 57	11 41	4 56		..
10 2	7 7	5 0½	60½	Portpatrick ......arr	10 7	11 50	5 5	..	..
..	..	..	..	Belfast—Steamer .. arr	Irish	time	9 0	9 0	..

† A Coach from Station to Gatehouse at 9-55 a.m. & 4-5 p.m.; also from Newton-Stewart to Wigtown and Whithorn at 3-30 p.m., on arrival of 2-5 p.m. train from Castle-Douglas, and 2-30 p.m. from Stranraer.

PORT-PATRICK & STRANRAER TO CASTLE-DOUGLAS, ETC.

FARES fr. Stranraer 1 Cl. s. d.	2 Cl. s. d.	3 Cl. s. d.	Miles.	Belfast—..Stm'r leaves	Irish	time	9 30	..	..
				Classes	1 2 3	1 2 3	1 2 3	1 2 3	..
				Trains leave	a. m.	a. m	p. m.	p. m	..
			..	Portpatrick.........at	..	10 15	1 10	6 15	..
..	..	..	..	Colfin...................	..	10 25	1 20	6 25	..
				Stranraer Station arr.	..	10 35	1 30	6 33	..
..	..	..	..	Stranraer Station lea	7 35	10 46	2 30	6 40	..
0 5	0 4	0 2½	2½	Castle Kennedy	7 43	10 52	2 37	6 48	..
1 0	0 8	0 5	5½	Dunragit	7 50	11 0	2 44	6 54	..
1 6	1 1	0 8½	8½	Glenluce	8 0	11 9	2 55	7 2	..
2 10	2 2	1 5	17	Kirkcowan	8 20	11 32	3 14	7 25	..
4 0	3 0	1 11	23½	Newton-Stewart †........	8 32	11 47	3 30	7 40	..
4 5	3 4	2 2½	26½	Palnure	8 40	11 56	3 38	..	..
5 0	3 9	2 5½	29½	Creetown	8 50	12 5	3 48	7 57	..
5 9	4 4	2 10	34½	Gatehouse Station	9 2	12 16	4 1	8 9	..
7 6	5 7	3 8½	44½	New Galloway	9 30	12 40	4 25	8 33	..
7 9	5 10	3 10½	46½	Parton	9 37	12 47	4 32	8 40	..
8 3	6 3	4 1½	49½	Crossmichael	9 47	12 57	4 42	8 50	..
9 0	6 8	4 5	53.	Castle-Douglas ....arr	9 58	1 5	4 50	8 58	..
..	..	..	..	Arrives at	a. m.	p. m.	p. m.	p. m.	..
..	..	..	..	Dumfries	11 0	2 0	5 47	9 49	..
..	..	..	..	Lockerbie	12 6	2 50	6 45	10 16	..
..	..	..	..	Glasgow v. G. & S. W...	8 55	8 50	10 15	..	..
..	..	..	..	Do. v. Cal.........	..	6 0	9 30	12 53	..
..	..	..	..	Edinburgh "	..	5 45	9 10	12 45	..
..	..	..	..	Carlisle v. G. & S. W...	12 28	3 46	7 28	12 28	..
..	..	..	..	Do. v. Cal.	12 53	4 45	7 35	12 37	..

† Coach from Gatehouse to Station at 7-30 a.m. and 1-30 p.m.; also from Whithorn at 5-15 and 8-30 a.m., to Wigtown at 6-45 & 10-30 a.m., via Garlieston, to Newton-Stewart, in connection with train.

Murray's Railway Guide passenger timetable for the line in 1867.

Mr F.W. Hutchinson.
Manager Portpatrick and Wigtownshire Joint Railways.

Chapter Four

Years of Uncertainty: 1861–1872

As already noticed, the Portpatrick Railway and the G&SWR were not on the best of terms when the new line was under construction. The latter, supposing it could drive a hard bargain, intimated that its terms for working the line would be the receipt of 72 per cent of the gross takings (which the Portpatrick Railway's Directors thought excessive), and further, as already mentioned, insisted that its own offer towards the cost of building the line was subject to a condition which was unlikely to be met. The Portpatrick Railway Board had never really reckoned on the Irish company putting down a further £20,000, since until Portpatrick harbour was able to take the packet steamers (which was not yet the case) the making of the line in Galloway would bring no advantage on the Irish side. It considered the G&SWR's attitude as unfair and dishonourable. So the Board determined to operate the line itself. The G&SWR representative then ceased to attend Board meetings, believing that the project would soon go bankrupt after which the G&SWR would be able to take it over.

They could not have been more mistaken. The Portpatrick Directors were determined to see the enterprise through, even though it meant further borrowing. True, so far as Galloway was concerned the new line could not be more than a considerable convenience to a comparatively small number of people, but once the packet boat service was in operation additional revenue from the Irish traffic would enable it to stand on its own feet. So, at any rate, it was hoped.

The G&SWR soon had other things to worry about, one of which, an ill wind in itself, eventually blew some good to the Portpatrick Railway. In 1862 the North British Railway opened the direct 'Waverley' route from Edinburgh to Carlisle by way of Galashiels and Hawick. In 1853 the G&SWR and the CR had each agreed to encourage and foster the other's traffic as much as possible. However, when the latter lost its monopoly of conveyance between Edinburgh and Carlisle, the former gave the NBR an undertaking that goods traffic from the southern part of its system should go to Edinburgh by the new route. Regarding this as a breach of the agreement, the CR sought an opportunity to retaliate, and this came when a line was proposed to link Lockerbie, on the CR main line, directly with Dumfries. This venture was promoted by local people in 1860, and the G&SWR at first ignored it, not supposing that it would ever be built. However, Parliamentary approval was gained, construction began, and the CR then stepped in and offered to work the line for as little as 30 per cent of the total receipts. The offer was accepted, and the new line opened in September 1863.

At this stage the Portpatrick Railway Board, which had had to borrow heavily to get its line built, announced that it would apply to Parliament for permission to issue another £20,000 worth of shares, to borrow a further £10,000, to make agreements with the CR and LNWR (over which latter company most of the through traffic from south of Carlisle would come) to run trains between Castle Douglas and Dumfries on to the Lockerbie line, and to raise money to buy steamships for the short sea passage to Ireland from either Portpatrick or Stranraer. The G&SWR now began to climb down its ladder and offered once more to buy £40,000 worth of Portpatrick Company shares, at the same time lowering its terms for running the line from 72 to 45 per cent of the receipts.

While the Portpatrick Company's Directors were digesting this, a counter-offer was received from the CR to subscribe the same amount and to work the line for only 43 per cent, the rate to be reduced further should traffic increase. At the same time the LNWR said it was ready to contribute a further £10,000. The G&SWR now descended a step further, offering the same terms as the CR – but it was too late; its rival's offer had been accepted.

This was not quite the end of the story. Many CR shareholders thought their company was going outside its proper sphere in undertaking to work a line in the far south-west of Scotland, and this view was strongly expressed at a shareholders' meeting. The result was a veering towards friendship with its rival on the part of the CR and a suggestion that they should work the Portpatrick line as partners – a solution that was to be adopted many years later. The two companies carried on a flirtation for some weeks over this matter. But the Portpatrick Railway, like Shylock, wanted to keep to the bond; its treatment by the G&SWR still rankled. Eventually the CR ceased to dally with its former foe and made a definite agreement, to last for 21 years, to work the Portpatrick line. This came into operation on 1st October, 1864.

Meanwhile the smaller line had been managing on its own with such engines and rolling stock as it had been able to acquire, for over 3½ years. It had had to operate on a shoe-string, but it had brought definite benefits to the area. More people were now travelling to, from or through Galloway than had previously travelled by road. The line's opening had brought visitors to local hotels, whose proprietors organised horse-omnibus links with the nearest railway stations. The products of Galloway farms were now being transported rapidly to distant places. Cattle no longer had to travel on the hoof; from the first summer of operation cattle trucks conveyed them, not only from Galloway but from Northern Ireland in special steamers to Stranraer, to destinations as far afield as East Anglia. Dairy products like butter and cheese arrived in England from Wigtownshire for the first time. There had been an improvement also in the local postal service; letters from England which had formerly not reached Stranraer till five in the afternoon were now arriving at mid-day.

Nevertheless the expected mail service to Northern Ireland had not yet materialised; the promised improvements at Portpatrick harbour had not yet begun. Stranraer harbour, however, *was* useable and the Portpatrick Company wished to begin a daily service at the earliest opportunity. It was reckoned that the passage to and from Larne could be made in a little over 2 hours. The Government, however, while procrastinating on the works at Portpatrick, was not yet favourably disposed towards using the Stranraer route for the carriage of mails, despite this being urged upon it by M.P.s from constituencies on both sides of the North Channel.

However, there was a potential traffic in goods and passengers which a regular service of steamers leaving and arriving at fixed times daily could hope to obtain. Early in 1862 the Portpatrick Railway's Directors had an encouraging meeting in London with representatives of other railways whose traffic might also be augmented if such a steamer service were instituted. The next thing to do was to find a suitable boat. The rail link between Belfast and Larne had now been completed and only the sea link was still lacking. No shipping company could be induced to supply a boat for the service, so one was purchased jointly by the companies which served both ports – an iron paddle steamer, the *Briton*, originally built to ply between Glasgow and Belfast, 175 ft long, of 24 ft beam, with a draught of 12 ft, a displacement of 492 tons, engines of 150 hp and accommodation fore and aft for over 400 passengers. Sailings began on 1st October, 1862, the advertised time of passage being 2 hours 45 minutes.

However, during three months of operation it arrived late a number of times, and storms sometimes prevented it from leaving port. Receipts during this period showed a loss since all the expected traffic did not materialise. The Directors of the Irish company thought a larger boat was needed, but the Portpatrick Railway Board, already losing money on its railway operations, felt it could not risk the expense. There was hope that before long the loss might be turned into a profit, since the trade in cattle was considerable. But it could not be denied that the *Briton* was far too small, though funds to replace it were not to be had. So on 31st December, 1862 the last voyage was made, and the ship was then sold to the Bristol Steam Navigation Company, which used it for 30 more years in journeys from the Bristol Channel to the south of Ireland.

The first steamship venture was a failure, but on the mainland the train service to and from Castle Douglas was kept going, and the CR's offer referred to above promised to get the line out of its financial difficulties. A marriage was arranged and shortly took place, with the G&SWR as the disappointed suitor.

The agreement between the CR and the Portpatrick Railway was approved, with some dissenting voices, at a General meeting of the former in November 1864, and unanimously at a similar meeting of the latter the following month, the satisfaction of the Portpatrick Company's shareholders being secured by the assurance that the LNWR had bought £10,000 worth of shares and that all outstanding debts had now been met, so that it would now be possible to pay a small dividend. The CR was given the right to work the line for 21 years from 1st October, 1864, taking over the whole of the plant, valued at over £37,000, which would be handed back on the expiry of the agreement in 1885. For working the line it was to receive between 38 and 33 per cent of the gross receipts on a sliding scale, depending on whether they were under £10 a mile, more than £15 10s. or somewhere in between, daily. This was about half of what the G&SWR had originally demanded. Most of the line's expenses, including the payment of staff wages and salaries, fell upon the CR.

From the smaller company's point of view it certainly seemed a bargain. Both parties hoped for profits. The CR took steps to re-insure its undertaking by coming to an arrangement with the LNWR, which agreed to receive one-third of the profits accruing yearly, or to bear one-third of the loss, whichever the case might be, in return for the right to work the line jointly with the CR should it wish to (which it never did) and for an assurance that all traffic passing from or to the Portpatrick line, to or from places south of Carlisle, should travel over the LNWR. Permission was obtained from the G&SWR to run trains over the Castle Douglas and Dumfries line and to use the G&SWR station at the latter place, in return for an annual rent of £75. Doubtless the news that the CR was having a survey made to ascertain whether a separate line of its own, linking the Lockerbie line with the Portpatrick one, was feasible, helped to secure this permission. No such line was ever built and it is questionable whether it was ever really intended.

Once in *de facto* possession of the Portpatrick Railway, the CR began to press for improvements such as the installation of more passing places and the building of company houses for employees. The Portpatrick Railway's Directors objected to the expense of such improvements, part of which they would themselves have to meet. Gradually the good feeling between the two companies degenerated into a grumbling relationship which endured over the whole 21 years. Four more passing places were eventually established, and a water tower was erected at Loch Skerrow for the benefit of thirsty locomotives which had climbed the bank from the Loch Ken viaduct. The CR also wished to introduce the train staff system along the line, instead of clearances for the passage of trains being given by telegraph, but they could not persuade the Portpatrick Company to make this admittedly desirable change. It also made alterations in train connections eastwards from Dumfries, which made some journeys impossible. Such prods and pinpricks were causing the happy union to degenerate into a series of tiffs. For all that, the Portpatrick Company had no option but to continue with the alliance, which had released it from its debts and enabled the payment of small dividends.

The CR was also anxious to re-establish a regular steamer service to Northern Ireland, and it persuaded the Portpatrick Board to co-operate, despite the latter's fingers having been burned over the failure of the *Briton* venture. The arrangement reached was for the smaller company to buy two vessels at a cost of £30,000, which the CR would operate, subject to an annual payment by the Portpatrick Company of £800 towards the expenses of working them. At the end of seven years the CR would refund the purchase money. The boats were to ply between Stranraer and Belfast. The service commenced at the beginning of December 1865 once the first of the ships had been bought. This was a paddle steamer, the *Fannie*, larger than the *Briton*, 231 ft in length, 26 ft across the beam, displacing 650 tons and having 250 hp engines. The *Wigtownshire Free Press* wrote appreciatively about it:

> Her accommodation for passengers, cattle and goods traffic is superb. Besides the usual cabins below she has a deck saloon, elegantly and comfortably fitted up. She has accommodation for 300 cattle, with the advantage of a gangway for the animals to walk to and from the 'tween decks instead of their requiring to be swung up and down as in most other cattle steamers. A number of horse stalls have also been carefully fitted up on deck.

Comfort, in fact, for all parties as they went on their lawful occasions – or to the slaughterhouse. A change, too, for the *Fannie*, which had spent its earlier years running the blockade during the American Civil War. A little later she was joined by a sister vessel, the *Alice*.

This second attempt to institute a daily passage to and from Ulster was as unsuccessful as the first, though it lasted longer. For one thing, it operated at times inconvenient to the public. Furthermore, an outbreak of cattle disease in 1866 led to the entire traffic in cattle having to be suspended for a while. Yet another setback was that the Post Office refused to send the mails by that route except on unacceptable terms. Some of the grievances felt locally were allayed when in August 1866 the trains between Euston and Stranraer were accelerated; even then, however, it took 16 hours to reach Belfast from London by the boat service, which did not arrive until 2 am the following day. The trouble was that the boats were slow. Criticisms began to be voiced at shareholders' meetings; in particular it was felt that one fast vessel would have been better than two slow ones. In the upshot it was decided to lay up the *Fannie* and use only the *Alice*, and the arrival and departure times at Belfast were altered to more convenient hours.

By now the CR was becoming disenchanted with its steamship venture, which was losing money. An almighty row was also brewing between its management and the shareholders over alleged financial mismanagement. It had been agreed that the boats should run for seven years, but the CR Board was increasingly ready to escape this obligation, and Nature gave it an excuse. In January 1868 the *Alice* was severely damaged in a storm and had to be sent for repair. The *Fannie* could not be used because its certificate of seaworthiness had been allowed to expire. So the service was suspended, and when both vessels were at last declared fit to sail they were instead sold, and that was the end of the second ferry service to Northern Ireland. The Portpatrick Board was indignant. It had agreed to subsidise the steamers on the understanding that the service would continue for 7 years. Now this undertaking had been prematurely broken. However, anger was tempered by the recognition that the CR was handicapped by the attitude of its shareholders, who had successfully insisted that an investigation committee should examine the company's financial position.

One consequence of this enquiry was that the CR again began to make overtures to the G&SWR about jointly managing the Portpatrick line. Things never got as far as that, but agreement was reached about the route to be taken by trains going to or from Stranraer. Those running east of Dumfries were to use the G&SWR line to Carlisle by way of Annan, a route seven miles shorter than that through Lockerbie, so that journey times were accordingly lessened. From the Portpatrick Railway's point of view, so far, so good. But the frequency of the service which its partner provided was still not good enough. In the summer of 1868 the provision of four passenger trains each way between Castle Douglas and Stranraer had been reduced to three.

To add to these irritations the CR demanded part of the compensation money paid by the Government when the harbour works at Portpatrick were abandoned in 1868, maintaining that the chief reason for their agreeing to work the Portpatrick Railway at such a low rate in 1864 was their expectation that the Portpatrick–Donaghadee passage would become a fact. But this was 'not in the bond'. All the CR was entitled to was a proportion of the traffic receipts; the compensation paid had been for carrying out works which had been unproductive, a different matter entirely. The money received would be used to benefit shareholders, of which the CR was one. This contention was legally correct and the CR had to swallow its disappointment.

At about this time, despite the cancellation of the improvements in Portpatrick Harbour, a private company was formed in the spring of 1868, the Donaghadee and Portpatrick Short Sea Steam Packet Company, to provide a rapid passage to Northern Ireland. The railways on the Scottish side agreed to authorise through bookings in connection with the sailings to and from Portpatrick. An iron paddle steamer, the *Dolphin*, somewhat smaller than the *Fannie* and *Alice* and very much older, began from mid-July to make morning and afternoon crossings from Donaghadee, returning at mid-day and during the evening. The enterprise got off to a good start, conveying many excursionists; during the first two weeks of operation between 100 and 200 passengers were carried daily. But difficulties soon began to appear. There were times when, in bad weather, the ship could not leave either port. An attempt to run passenger trains down to Portpatrick Harbour failed because the junction was a trailing one just beyond the station platform, with only a short shunting neck beyond it, so that only two or three 4-wheeled coaches at a time could be manoeuvred down to the quayside. From 21st September there was only one sailing a day each way, with an accompanying increase in fares. Then the *Dolphin* had a slight accident, which put it out of action. And that was that. The Short Sea Company at once sold the vessel. It had lost £2,500 and was not prepared to take any more risks. The fact was that both harbours were unsuitable for regular traffic in all seasons, and the coast of the Rhinns was in any case subject to fierce winter storms.

Despite the failure of three successive ventures to set up a cross-channel service, the nearness of the Irish coast and Belfast ensured that the possibility was kept in mind. It was in everybody's interest to overcome difficulties and discussions between all the five companies concerned, including the LNWR, were held. The spokesman for the Portpatrick Railway, Lord Stair, reported periodically to the shareholders. While the CR, the G&SWR, the Portpatrick and the Belfast and Northern Counties Railways were all directly interested in establishing the steamer service, the LNWR had mixed feelings. What it might gain on this one swing had to be set off on as many as three roundabouts – the other three steamship companies which also plied to Ireland and in which the LNWR had invested. However, eventually agreement was reached. The Portpatrick Company was to contribute £10,000 towards purchasing a new steamer, the contract for which was placed with Tod and McGregor of Glasgow in August 1871.

Nine months later the ship was launched – a paddle steamer like its predecessors, 211 ft long, 24 ft in beam, 13 ft in draught, with a displacment of 500 tons and engines developing 200 hp. 500 passengers could be comfortably accommodated, those in the first class saloon enjoying a measure of luxury. It cost £18,500 and was named *Princess Louise* after a daughter of

Queen Victoria who had recently married the Marquess of Lorn. Weekly sailings began on 1st July, 1872, out to Larne in the late morning and back to Stranraer in the early evening. Her speed did not at first meet expectations, but after some attention by the builders to the engines she proved able to make the passage in 2½ hours. Thus one of the aims of the Portpatrick Company's promoters was at last realised; their line was no longer just a provincial backwater. It only remained for the Post Office to despatch their mails to Northern Ireland by this route, but it, cautious as ever, waited before making up its mind.

However, one concession was wrung out of the Government, which was persuaded to contribute more towards the cost of running the mail trains into and out of Galloway. The Post Office agreed to augment the amount it paid, providing additional trains were put on to improve the local mail service, and that a sorting van was included in the chief train from Carlisle. A figure of £1,154 per annum was proposed and accepted. The consequence was not only that the mails arrived sooner, but also that there would now be four trains running the length of the line in either direction, so that one of the Portpatrick Company's grievances against the CR was removed.

10th June, 1904 at Stranraer Harbour. *Oakwood Collection*

Chapter Five

Dividends and Disputes: 1872–1885

After the CR had taken over the working of the Portpatrick Railway there was a modest improvement in the latter's fortunes, with yearly dividends, small though they were, being declared. After 1872 matters improved still further. In part this was due to an improvement in national prosperity which had begun a little earlier. Agriculture and commerce, assisted by the nation-wide extension of the railway system, were beginning to flourish, and the railways shared in the general prosperity through the demands made on their services. The CR emerged from its financial troubles and its shareholders ceased to needle it. One reason for its own and the Portpatrick Company's better prospects was of course the establishment of the Stranraer–Larne ferry service. People were now not only travelling to and from Galloway, but through it.

A further advantage to both companies was the belated agreement of the Post Office, in 1875, to send mails to Northern Ireland by the Stranraer route. At three-tenths of a penny for each letter carried the Post Office's terms were not exactly generous, but it all added up. In consequence there was an improvement in the Portpatrick Company's annual dividend, which rose from 2 per cent in 1872 to 4¼ per cent in 1877, dipped to 2¼ in 1879, rose to as much as 4½ in 1883 and then fell back to 3¼ in 1885, the year when the agreement with the CR ended. Not all the surplus money went to the shareholders, however. In 1874 a loan was made to the Larne and Stranraer Steamboat Company to meet a temporary shortage of funds; this was quickly repaid. The next year a new vessel to complement the *Princess Louise*, the *Princess Beatrice*, of much the same size and power, was launched from Harland and Wolff's Belfast yard. The *Louise* was then laid up for a while to be fitted with new boilers.

A 19th century railway was both a public service and a business concern, and the convenience of travellers had to be weighed against the need to keep shareholders happy. In the case of the Portpatrick Railway, as with other lines, these two *desiderata* did not always balance out to everyone's approval. Now that the link with Northern Ireland had been established the Company had conflicting claims upon it, the needs of passengers travelling from and to destinations well beyond the sphere of local operations (which meant, ideally, the provision of fast trains to and from Stranraer, stopping as infrequently as possible) and the needs of local residents (which meant trains stopping as often as possible at times and places convenient for local needs, especially on market days). Many complaints appeared in the local press, for the railway was still the only convenient means of making a journey of any length; road vehicles were all horse-drawn and very slow. The local traveller, it seemed, had only Hobson's choice. However, even worms will turn, and when repeated complaints seemed to be having no effect a tendency appeared for groups of persons to hire road vehicles in which to travel together, particularly on market days. The Portpatrick and Caledonian companies then became wise to what was happening and put on extra trains.

While the two companies were now on fairly good terms, relations between the CR and the G&SWR all too easily became strained, and each could be deliberately uncooperative towards the other, doing itself no good in the process and inconveniencing the travelling public. In September 1872, for example, there was an occasion when two trains, a CR one routed by way of Lockerbie and a G&SWR one going via Annan, both bound for Dumfries to connect with a train for Stranraer, were ready to leave Carlisle at the same time. The CR officials refused to let the other company's train start until its own was sufficiently far ahead to be able to arrive at Dumfries first. However, what one company could do at one station the other could do at the other, and the Stranraer train was deliberately delayed at Dumfries until the one routed through Annan had arrived to connect with it. This sort of tit-for-tat behaviour frequently kept the steamer waiting at Stranraer. There was also a good deal of late running, allegedly due to insufficient locomotive power, and, more seriously, a few accidents, as described in a subsequent chapter.

During the remainder of its independent existence the Portpatrick Company suffered a grievous loss in the early death of its vigorous and diligent Secretary, Alexander Ingram, who had been prominent not only as the line's chief executive officer in control of its day to day operations but also as a local townsman. In the Burgh of Stranraer he was at various times Clerk to the Peace, Clerk to the Circuit Court, magistrate, Chairman of the School Board and, for the last five years of his life, Provost. Tactful and popular, knowing everyone locally, he packed the activities of a longer lifetime into a short span of years and died in 1875 at the age of 48. He went from the scene just as his skills and persuasiveness were most needed, since in that year his Company was in dispute with the Burgh Council over the pier used by its boat trains and by the steamers with which they connected.

The East Pier had been built and was maintained by the Council and the railway paid an annual rent to use it which covered all the dues that would otherwise have been payable. The pier's nearer and further ends had been built of stone, but the middle part, less substantial than the rest, was of wood. Until 1873 it was not considered safe to take trains along the pier; they stopped short at the stone approach and passengers were obliged to walk to or from the ships along an unsheltered path. After that date a small engine of light weight was kept specially at Stranraer Town station to haul carriages along the pier towards the boat so that passengers did not have so far to walk. But the pier was still lacking in shelter, and two years later was also found to be badly in need of structural repair. The Portpatrick Company asked the Burgh Council for this to be done immediately, saying further that if it were not done they would do the repairs themselves and deduct the cost from their annual payment. This request, couched in an un-conciliatory manner, provoked the Council to reply that it needed a larger annual payment to meet the costs of upkeep. The Company then replied that the Council was bound by its existing undertakings, and said that if nothing were done it would complain to the Board of Trade.

Nothing *was* done, so the Company proposed to take over the pier through an Act of Parliament. The Council opposed this, and the dispute was carried as far as the House of Lords. In the end the two parties were persuaded to resolve their dispute amicably. A new agreement was drawn up, by which the Council was to receive £500 annually but was no longer required to maintain the pier, the railway undertaking this duty. The Portpatrick Company also paid the legal costs of the dispute. However, the whole affair caused much bitterness, and the company also provided a rod for its own back by arranging to share the management of the pier with the recently-formed Girvan and Portpatrick Junction Railway, whose origins and involvement with the Portpatrick line are described below. The whole business had cost £8,000, most of which could have been avoided if there had been more mutual good will. The pier itself was repaired, strengthened and provided with sheltered accommodation for waiting passengers, an asphalted approach road and a powerful crane. All was finished by 1880.

Other important events during the later 1870s were the construction of two more railway lines in the region and the projected building of a third. To take the latter first: during 1877 M.J. Stewart, M.P. for the Stewartry of Kirkcudbright and one of the Portpatrick Railway's Directors, who lived at Ardwell in the Rhinns, proposed the building of a line from Stranraer southwards along the eastern side of the peninsula beside Luce Bay for some 13 miles to the vicinity of his own residence, with the possibility of its being extended later either to Port Logan or Drummore, and he employed a surveyor to plan its route and estimate its cost. He managed to interest his fellow-Directors. It looked likely to be a profitable venture. The areas into which it would reach had considerable potential for exploitation as a holiday resort, and the local farms produced quantities of milk which, rushed to Glasgow by train, would find a ready sale there. The project might well have succeeded had it not been for a temporary trade recession which deterred people from promising to buy shares. For some five years the promoters remained sanguine, but eventually accepted that the scheme would not get off the ground, and the undertaking was abandoned.

Of the two lines in this region that *were* actually built, one, the Wigtownshire Railway, is the subject of the next chapter. The other was the Girvan and Portpatrick Junction Railway, for whose construction an Act of Parliament was obtained as early as 1865. There was much local feeling in favour of a railway that would provide a direct link between Glasgow and Stranraer. The existing route by way of Dumfries was 156 miles long; by way of Girvan and Ayr it would only be about 100, so economies in times and fares would certainly ensue. The people of Stranraer would indeed benefit, but for the Portpatrick Railway and the CR the prospect of losing traffic to the shorter route was unwelcome. The latter's Board knew that, if built, it would almost certainly be worked by the G&SWR, and was particularly hostile. Some traffic, indeed, might be gained, since the new line was planned to connect with the Portpatrick line six miles east of Stranraer, so some of the receipts would need to be shared, and of course the gain would be greater if additional traffic were attracted to the Stranraer–Portpatrick branch as well. But, all things considered, the interloper looked like eroding rather than increasing revenue. Nevertheless, a few of the Portpatrick Directors also agreed to serve on the Board of the new line.

In this confused situation it came as a relief to most parties of the existing establishment when the Girvan and Portpatrick Junction project was shelved in 1867. Enthusiasm for it had been considerable in Stranraer, but there was little support in other places along the proposed route. Had the latter run as originally suggested, through Ballantrae and along the shore of Loch Ryan it would have served more people and commanded more support. However, to have taken it that way would have entailed a far larger first cost, and when a route across the moorlands to the east, taking in only a handful of small villages, was suggested instead, few were willing to invest in a railway which for a great part of its course would traverse an almost uninhabited area. Eventually its promoters accepted that it could not be built for some years.

In 1871 things began to look brighter; increased national prosperity induced the purchase of shares, and the actual construction of the line was taken in hand. The Portpatrick Company accepted the inevitable with as good a grace as possible. Then something happened to sour their attitude. While negotiations were in progress about the terms on which trains would be allowed to use the existing line from the proposed junction near Challoch Farm to Stranraer, it became known that the new company was engaged in having an alternative route surveyed by which it might reach Stranraer entirely on its own metals. Were this to happen the existing line would get no good at all from the new line, but instead lose traffic to it. The Portpatrick Directors, with some dissentients, decided to oppose the Bill in Parliament. While the latter was considering their objections, however, a deal was struck. The idea of a separate entry into Stranraer was given up and terms were agreed for the use by both parties of the existing line between Challoch Junction and Stranraer.

The new line was completed in 1877, after very great difficulties, but the Portpatrick Railway and the CR had now once more become unfriendly to it and deliberately made difficulties about their track and stations being used. Consequently the first trains from the Girvan direction had to stop six miles short of Stranraer and send their passengers on by chartered horse buses. The Portpatrick Railway and the CR resorted to many tricks and stratagems to prevent the intended services on the new line from being implemented. The Board of the Girvan and Portpatrick Junction Railway now attempted to raise more money. By 1877 it was in considerable difficulties. The line had cost far more to build than expected and was failing to attract traffic from along its route; it had had to conclude an unfavourable agreement with the G&SWR for working it; the earlier period of national prosperity had been followed by a slump; steamer companies were undercutting it with their services from the Clyde to Belfast. It had many creditors, who in 1879 succeeded in getting it put into the hands of a judicial factor, since it was now so deeply in the red.

The question most at issue was how payment for the upkeep of the jointly-used section from Challoch Junction to Stranraer should be apportioned. The matter eventually went to arbitration and was settled by a decision of Sir Thomas Bouch, whose award in June 1880 seems very fair. Generally the new company was to pay either half or two-fifths or one-third of the maintenance or improvement expenses under eight different headings. However, the Girvan and Portpatrick Junction Company was then in no position to pay anything at all, and by the summer of 1881 it owed the Portpatrick Company over £10,000. The latter then began proceedings in the courts to recover its money. The response of the smaller company was to seek from Parliament authorisation to change its name and to borrow £30,000 with which to pay off its immediate debts and buy locomotives and rolling stock to enable it to work the line itself. This was granted in 1882. It then offered to mortgage itself to the Portpatrick Company but the latter rejected the suggestion. It seems very likely that the Portpatrick Company was hoping it would soon be able to buy up a bankrupt concern and manage it itself. But the smaller company struggled gamely on, and at length managed to negotiate a more satisfactory working agreement with the G&SWR. In consequence by September 1883 its debts were all paid off, it had re-possessed its line and was at last allowed to run its trains into Stranraer. An unhappy conflict had ended with a negotiated peace.

And so the critical date approached when the 21 years' agreement with the Caledonian was to end. 1885 was to see unexpected changes, a reconciliation between two enemies of long standing, and the disappearance of the Portpatrick Railway as an independent company. Meanwhile one must turn back a few years to the formation of the Wigtownshire Railway and the intriguing tale of its early fortunes before it too lost its separate existence in the same merger that absorbed its neighbour.

The abutments of a bridge which formerly carried the harbour branch over the main road in Portpatrick. This branch was closed about 1870. *D.S.M. Barrie*

Chapter Six

The Wigtownshire Railway

The triangular southward extension of the County of Wigtown known as the Machars, whose base is a line drawn from the estuary of the Cree westwards to the mouth of the Water of Luce, more or less corresponding with the line of the railway from Newton Stewart westwards to Glenluce viaduct, is its most fertile and productive region. Along its shores are the burgh and port of Wigtown which gave the shire its name, two other small harbours, Garlieston and Isle of Whithorn, which face east and south respectively, and the pleasant little watering place of Port William half way down the south-west-facing coastline which looks towards the Mull of Galloway. Inland it is a region of farms and villages with some bare moorland patches. Near its southern tip is the small but historic township of Whithorn, where Saint Ninian's first Christian monastic settlement was made during the 5th century. In the early 19th century a growing trade had developed between the Machars and ports on the Irish Sea and Clyde coast, particularly Liverpool and Glasgow, and small cargo steamers called at Wigtown, Garlieston and Isle of Whithorn, so that the region's agricultural and dairy produce found markets a considerable distance away.

As soon as the Portpatrick Railway had begun to operate, the making of a branch line penetrating the Machars came under discussion locally. The railway's Directors were much in favour of such a valuable potential feeder to their own system, and its Chairman, Lord Dalyrymple, toured the area in 1862, along with other officials of his railway, including its Consultant Engineer Mr Blyth, to see if they could generate any interest. In January 1863, at a public meeting in Wigtown, Blyth expressed his belief that a railway could be built from Newton Stewart to Wigtown for £20,000; to take it further to Sorbie would raise the cost to £60,000 and a further extension to Whithorn would raise the total to £85,000. Sanguine forecasts – far too hopeful, as things turned out – were made as to its profitability, and the Burgh Council of Wigtown offered to contribute £2,000. A provisional committee was then established to consider the project, and a survey of a suitable route for the line was made by Mr (later Sir Thomas) Bouch. However, no further steps were then taken. The Portpatrick Railway was finding it could barely pay its own way and could not offer more than moral support. Not until another 8 years had elapsed, and prosperity seemed to be returning everywhere, did the proposal surface again.

This occurred in September 1871 when a public meeting to commend the enterprise was held in Newton Stewart. There were many expressions of enthusiasm for the proposed line, and a fresh committee was formed to discuss what might be the best route for it to follow. When this met, two views were put forward. One, the majority view, favoured a line running south from Newton Stewart to Wigtown, thence veering inland to Kirkinner and Whauphill, turning south-eastwards through Sorbie to within a short distance of Garlieston, and finally heading southwards to Whithorn. The other preference was for a branch to leave the Portpatrick line at Kirkcowan, seven miles west of Newton Stewart, which would be directed more or less southwards, going nowhere near Wigtown but eventually converging upon the other route at Whauphill, Under this proposal Wigtown would be served by a separate branch from Newton Stewart, so that there would then in effect be two Wigtownshire Railways. Sir William Dunbar, the M.P. for Wigtown Burghs, who was the chief supporter of the second suggestion, was in fact envisaging a branch which could be extended from Kirkcowan to the north. The Girvan and Portpatrick Junction Railway had already been authorised and its construction was soon to begin, and Dunbar's idea was for the Wigtownshire Railway to extend backwards to make a junction with it, so that a short through route to Glasgow from the Machars could be established by way of Girvan and Ayr, which would benefit the whole county. However, he did not carry his point, the meeting approving the route through Wigtown by a large majority. Sub-committees were now formed to canvass for promises of subscriptions from the parishes of the Machars. At a second meeting of the full committee another attempt by Sir William Dunbar to get the route altered was rejected, and a third committee was appointed, entrusted with the task of getting the Bill through Parliament. A London engineer, Mr. Tolmé, was selected as a consultant for the construction of the line, and the name of the proposed company was fixed upon; it was to be called the Wigtownshire Railway.

The Bill proposed a line more or less along the route that Bouch had recommended 10 years earlier, from Newton Stewart to Whithorn, with a short tramway-branch to Garlieston harbour. It passed through Parliament without difficulty and became an Act in July 1872. The first meeting of the company's Directors was held six weeks later. Promises of financial support were now coming in. In April 1873 John Granger of Aberdeen was given the contract for building the line, and three months later his men began work on the excavations and trackbed. In 1874 a start was made with laying the permanent way from Newton Stewart, and by July this was ready for use as far as Wigtown, where a temporary station was put up pending the erection of permanent buildings later. It was confidently hoped that trains would soon be worked there from Newton Stewart.

It was taken for granted that the Caledonian Railway would work the line, as it had agreed to work the Portpatrick Line, and on similar terms. A formal request was made for it to do so, and in September 1871 the CR Board replied that it was prepared to undertake the task for as long as the agreement with the Portpatrick Railway lasted, but *not* on the same terms. Instead they would do it as 'cost price' – that is, whatever it cost to run the line would be chargeable to the Wigtownshire Company, irrespective of what the latter's takings might be. This of course meant that it intended to run no financial risk, while at the same time profiting from whatever additional revenue might come its way from passenger or goods traffic originating on or proceeding to the new line and passing over its own lines subsequently or beforehand. This did not suit the Wigtownshire Directors, and haggling now began between the two parties.

The Wigtownshire Company wanted a percentage arrangement. 'Cost price' was only too likely to lead to its losing money. A long drawn–out correspondence ensued. On the side of the CR, while there was a willingness to see the new line operating since it would serve as a feeder both to the

Portpatrick line and itself, it was equally determined not to be bound a second time by an agreement which proved in the end to be unfavourable. It was willing, however, to discuss the use of a percentage arrangement and a sliding scale, and negotiations went on for a long while, with the Wigtownshire Company at first entertaining CR proposals and afterwards having second thoughts. Eventually, in July 1873, the broad pattern of a working agreement was decided, on a sliding scale basis. The CR would receive 80 per cent of the gross traffic receipts if they did not exceed £5 a mile each week, but the figure would fall if they did exceed that figure to a lower limit of 50 per cent if the takings rose to £50–£60 weekly. On such a basis, the CR thought, it would find the undertaking profitable, though it would not be in pocket if the line were only lightly used.

Discussion now turned to a consideration of details. The Wigtownshire Company wished to avoid the expense of installing turntables, and a three-cornered exchange of letters followed between itself, the CR and the Board of Trade. The intention was to employ tank engines which would not require turning, but the CR did not wish to bind itself to doing this if it were allowable for engines travelling tender-first to be used when no tank engine was available. As to this, the Board of Trade eventually consented to relax the rule. All seemed to be going well and an agreement appeared imminent.

Then, at the last moment, negotiations broke down. It seemed to the smaller company that, as the line was to be worked by the CR, using its own rolling stock and locomotives and providing its own staff, the latter ought to be wholly responsible in case of loss or damage attributable to 'negligence, carelessness, culpability or want of skill' on the part of the employees. The CR jibbed at this and submitted alternative proposals, which the Wigtownshire Directors found unacceptable. More argument ensued, and it appeared that each side had reached a sticking-point beyond which it felt it could not go. In mid-September 1873 the smaller company regretfully decided that it would have to work the line itself, and told the CR so. Thus ended the protracted negotiations between the lesser party, which was determined not to be buried under a mountain of debt, and the larger one which was equally determined not to buy a pig in a poke.

The Wigtownshire Directors now faced the prospect of working the line themselves, with the precedent before them of the Portpatrick Company's unsuccessful experience during 1861–1864. They managed to get the CR to agree that the former's traffic superintendent, William Grafton, whose expertise they valued, should act also as the traffic superintendent on their own line, and they envisaged the initial provision of a service of two trains a day each way, at first to and from Wigtown and later along the whole line when it had reached its full extension southwards. Hampered by lack of funds, they knew they had to run the line as cheaply as they could, and even considered hiring locomotives and rolling stock instead of buying them. They estimated their minimum needs as two locomotives, six passenger coaches, ten goods wagons and a brake van. They made many enquiries but no offers came their way.

Meanwhile the line was being extended southwards. A girder bridge of two spans was built across the river Bladnoch, here a wide stream broadening into a tidal estuary south of Wigtown, and construction proceeded as far as Sorbie, the track reaching a summit at Whauphill up very steep gradients and then falling less steeply. Money was now beginning to run out, and at the end of 1874 a special meeting of shareholders authorised the company to borrow a further £20,000. This loan having been raised, a further extension to a point near Garlieston was begun, from which the intended tramway could run to the harbour. But no rails had yet been laid beyond Wigtown, and no trains were travelling even to that place, for no locomotives or coaches had yet been obtained. Something like a stalemate seemed to have been reached, when an unexpected benefactor appeared from a most unlikely quarter.

A letter was received by the Board from a Mr Thomas Wheatley, who, had it not been for certain untoward circumstances, would certainly never have dreamed of applying to it. One does not willingly step down from the position of locomotive superintendent on one of Scotland's chief railways in order to manage a small concern like the Wigtownshire line. However, Mr Wheatley was in his employer's bad books. He had been in trouble before. Having entered railway service in his youth on the Leeds and Selby Railway, he had then moved to the Midland where he had been employed as a driver. On a certain occasion he had handled a light engine so inexpertly when proceeding to assist another train that he had run into the back of the latter, and a passenger had been killed. At the Coroner's inquest a verdict of manslaughter had been given against him, though this does not seem to have been followed by any prosecution. However, he did not stay to face any charges, but simply absconded, and some time later found work on the Manchester, Sheffield and Lincolnshire line, eventually being put in charge of the locomotive depot at Grimsby. Subsequently he moved to the LNWR. Finally, when the position of locomotive superintendent on the NBR became vacant he applied for it and was appointed.

His work on that railway seems to have given professional satisfaction. His chief claim to fame while on the NBR was to build a number of inside-cylindered inside-framed 4–4–0s with front bogies, the first of this kind to be built in Great Britain, which set a trend that every major British railway was to follow. One of these locomotives, No. 224, fell with its train into the Tay estuary in January 1879 when the first Tay bridge was destroyed in a storm; it was later extricated, repaired and not only put back into service but twice rebuilt.

However, misfortune dogged Wheatley – or perhaps his ethical sense was less pronounced than his engineering skill. He lived in a house near Glasgow which belonged to the NBR, and was accused of furnishing and decorating it at the company's expense. He was also held responsible for the unaccountable disappearance of material from the Company's works at Cowlairs. He resigned under pressure and had to look for another job. It was brought to his notice that the Wigtownshire Railway needed someone to manage it, and after making a journey to Wigtown to see for himself what the possibilities might be he wrote formally to the Directors offering to work the line for 65 per cent of the gross receipts, for a period of 5 years. He was now

55, and perhaps hoped to pick up something better later on; as things turned out he remained at Wigtown until his death.

The Wigtownshire Board must have felt surprise at the windfall that had come their way, and no doubt took up references with the NBR, which seem to have been satisfactorily answered; it was, after all, a matter of indifference to the other railway whom he might work for, so long as it was not for them. In fact Wheatley was to prove a diligent and reliable manager, working the line to the Board's satisfaction, despite the fact that he had to live on a reduced income and did not find it easy to make ends meet. He gained a reputation locally for being friendly and obliging, helping farmers in the area in his spare time when some blacksmith's work needed doing. He ran the line in association with his son, William T. Wheatley, and by stringent economies kept it going.

In 1880 his 5 years' tenure ended, and was renewed after something of a tussle with the Board, when he consented to receive 62½ per cent of the total receipts, a little less than before. Less than 3 years later he died suddenly at the age of 62. His son asked to be allowed to continue to manage the line, and this was agreed to. But he was not to enjoy a long tenure, for very soon afterwards, as recounted in the following chapter, the Wigtownshire Railway's independent existence came to an end.

During the elder Wheatley's period as manager the line was finally extended to Whithorn and reached its full extent of 19.1 miles. This had not happened earlier because of shortage of funds and further borrowing was necessary to make it possible. A new Act of Parliament was obtained which enabled the issue of further shares and the raising of another loan; construction began in the spring of 1876 and Whithorn was reached 14 months later. The original intention had been for a light tramway to connect Garlieston with the nearest point on the line, but this was altered in the Act; a proper branch line was built so that there could be through traffic down to Garlieston quay, and for a short while, until Whithorn was actually reached, Garlieston was the actual terminal. The finished extension was ceremonially opened on 7th July, 1877, when the Board of Trade inspector, Major-General Hutchinson, rode the whole length on the footplate of a CR locomotive borrowed for the occason, and a celebratory meal was held in the Town Hall at Whithorn, where many speeches were made, including one by the Earl of Galloway, who uttered a hopeful forecast that a branch might eventually be constructed to link Whauphill with Port William. However, this never came to pass, and the people of Port William had to be content with a road conveyance to carry them six miles to the railway.

Chapter Seven

Under new management: 1885–1892

The prospect of being hanged, Dr Johnson once assured his friend James Boswell, concentrates the mind wonderfully. The prospect of losing the services of the CR as its working partner in 1885, when the existing agreement was due to expire, similarly concentrated the attention of the Portpatrick Railway's Directors as the fatal moment approached. However, it was not independent survival that the company desired, but suicide on favourable terms. As a business entity it was content to be wound up so long as its shareholders did not suffer. The Board had learned two things from past experience; it would not be able to work the line profitably if it tried to go it alone; futhermore, the CR regretted the generous terms it had agreed to for working it in 1864 and would drive a much stiffer bargain once the chance came.

However it was unthinkable that the line should not continue as a going concern when so much capital had been invested in it and so much passenger, freight and Post Office traffic now passed over it. The obvious thing to do was to sell it to the highest bidder, if one could be found to bid at all. As early as 1877, when the dividend level reached its peak of 4½ per cent, this was already being considered. Four and a half years later the first step was taken. The Earl of Stair, the company's Chairman, approached Richard Moon, Chairman of the LNWR, to find out on what terms that company and its Scottish partner the CR would be ready to buy the line outright. A year later, in 1883, Moon was discussing the matter with his opposite numbers on the MR and G&SWR; at the same time the CR agreed that an engineer from the MR should examine the Portpatrick line and estimate its condition. It looked as though a quadripartite agreement, for all four railways to run the line as partners, was imminent. The Portpatrick Railway Directors rubbed their hands in anticipation.

However, the chicken had not yet been hatched, and indeed nearly died within the egg. The four railway Boards, when it came to discussing details, and in particular who should pay how much to whom, began to disagree. While the LNWR and the CR were prepared to accept the existing terms by which running powers were granted by the G&SWR over its line between Dumfries and Castle Douglas, the latter, with MR support, wanted to charge more. Negotiations went on during December 1883 but no consensus was achieved. In the meantime the Portpatrick Company had lodged a Parliamentary Bill at Westminster, but it began to look as if it would have to be withdrawn, and Moon informed Stair about this. The Portpatrick line's Secretary wrote to the G&SWR urging that agreement be reached; the latter's reply showed that it did not mean to budge from its position. So in the middle of March 1884 the Bill was withdrawn and everybody was back to square one.

The Board of the Portpatrick Railway felt thoroughly let down, and Moon was informed that it intended to seek an agreement by which the MR and G&SWR would work the line together, and that if this approach failed the company would itself attempt to work the line. It also indicated that it intended to claim compensation for the expense it had been put to in promoting the Bill. The Directors certainly had cause to feel gloomy. So far they had been able for several years to distribute a reasonably good dividend. This, however, had been largely due to the low charge the CR had made for running the line – and regretted having made. With memories of the exorbitant terms the G&SWR had proposed when the line was being built, the Portpatrick Directors could not have had much hope that more reasonable terms would be accorded now. As to working the line themselves, they had tried to do that before without success.

It is true that not all members of the Board were so pessimistic. One, James Hosack, quoting the examples of other small lines that were managing on their own, thought the Portpatrick Railway might well emulate them and be able to offer dividends as good as theirs. He also suggested that another route be considered in order to by-pass the Castle Douglas and Dumfries line, avoiding the former town altogether. He did not succeed in convincing the other Board members, but a survey for a direct line between Parton and Dumfries was in fact made at this time.

Meanwhile the Wigtownshire Railway was in a position similar to that of the Portpatrick Railway; it was ready to be taken over for its own good. It did not have to face a deadline from a working company, since it was being operated, first by Thomas Wheatley, then by his son. Until November 1883, however, its Board knew nothing of the negotiations with the Big Four. In that month the Portpatrick line's application for a Parliamentary Bill became known. The Wigtownshire Board at once took steps to contact the same four railway companies and press the merits of its own case. Their line, they claimed, was potentially a profit-making concern well worth purchasing. Since its opening, steamers for Glasgow and Liverpool from Wigtownshire ports had almost ceased to operate; the railway now carried nearly all the goods that had previously gone by sea. William McClure, the company's Secretary, made a point of visiting both the CR and G&SWR General Managers to press the case for a take-over.

A rather unedifying dispute now broke out between the Boards of the two smaller companies over the extent to which the Portpatrick line had failed to gain assurances of a faster train service after the line had been disposed of, and about the advantages which the Wigtownshire Railway had brought to it. The Earl of Galloway was on both Boards, and represented the Wigtownshire's interests to the other Directors. The latter protested that in any case they could not run faster trains over a single line, and that the advantages the Wigtownshire line brought were nothing like as great as the latter supposed. They were still going at it hammer and tongs when the stage was removed from beneath them by the withdrawal of the Bill in March 1884.

Soon afterwards, however, things began to get moving again, thanks to Moon. He proposed to the CR Board that his company and theirs might work the Portpatrick line together on the existing terms, with the receipts due to them being mutually divided according to the proportions of traffic which had accrued to either. The CR agreed to this – which on the face of it seems surprising since under the proposed scheme they stood to get less money, not more. But Moon was a skilful man of business and understood the gentle art of bluffing, and this the CR presumably knew. No sooner had news of the

The locomotive which hauled the first train to arrive in Wigtown – a 2–2–2 well-tank engine with 5 ft 6 in. driving wheels, bought from the NBR (originally built in 1856). The date was 2nd March, 1875, the train comprised a single CR coach carrying the inspection party. *Author's Collection*

Wigtownshire Railway engine No. 5, a 0–6–0 well-tank seen here at Garlieston in 1890. *Author's Collection*

proposal got around than the MR and the G&SWR had a change of heart, and the four-sided negotiations which had been broken off began once more. In September 1884 the Portpatrick Directors were able to report to their shareholders that a new Bill would be placed before Parliament at its next session, and that they were sure it would go through successfully.

This time the Wigtownshire Railway had no reason to feel it was going to be left out in the cold. The CR sent a representative to examine the track of the railway in company with its Secretary McClure. The viaduct across the Bladnoch was given a careful inspection. On 28th September, 1884 a special party of CR Directors travelled the whole line in both directions. Meanwhile in London the Wigtownshire Company had secured the services of a barrister, Mr S. Swarbrick, to represent its interests, and the latter received an inventory of the plant and rolling stock on the line, which included the information that it had six locomotives, two tenders (the other engines being tank engines), 14 passenger coaches, 3 horse boxes and 31 goods wagons – not a large quantity and very mixed in quality, nearly all having been obtained second-hand.

Negotiations with the Big Four went on during October, and eventually a Bill was laid before Parliament that included both lines; it became law the following year. The Portpatrick Railway shareholders did rather better than those of the other line, as the new shares issued to them were equal in value to the ones they held, whereas each Wigtownshire share was only exchanged for half its nominal value. However, dividends were now guaranteed at 3½ per cent for the first company and half that amount for the second. The management of both lines was given to a Joint Committee which had representatives from each of the Big Four. It remained to be seen how well the new organisation would perform.

One interesting possibility mooted at this time, which never came to pass, was a southward extension of the Wigtownshire line to Isle of Whithorn. The harbour of this small fishing village was offered for sale to the CR and G&SWR companies in February 1885 by the Burgh of Whithorn which owned it. However, after the railway reached Garlieston the vessels that originally called at Isle of Whithorn now went instead to the former port, since transporting cargoes inland was now easier from there. The result had been that the Burgh went bankrupt, and it now wished to sell the harbour to meet its liabilities. But the railways declined the offer. For the port to have been of use to them a five miles' extension of the existing line would have been necessary as well as a rebuilding of the pier. If ships were using Garlieston harbour, why concern themselves about the other one? So nothing was done. One feels, however, that an opportunity was missed. Isle of Whithorn is only 26 miles from Ramsey, the Isle of Man's most northerly port. Had the five extra miles of line been constructed it would have been possible to run a passenger steamer service linked by rail with the populous regions of Central Scotland by way of Ayr and Girvan; spur lines at Challoch Junction and Newton Stewart would have made reversal of trains unnecessary. However, holiday traffic on the present-day scale did not then exist and could scarcely have been anticipated. As things are, Isle of Whithorn remains a quiet little backwater from which occasional summer excursions are run to the Isle of Man in motor boats for day trippers.

It was a sign of the new order of things that the first two meetings of the new Joint Committee were not held in Galloway but at Carlisle and Euston. They were called to sort out and correct anomalies and arrange for the immediate future. The first, on 25th August, 1885, had a high proportion of top brass present; the eight Committee members included the Chairmen of the LNWR, MR and CR as well as the Earl of Galloway who now represented the G&SWR. It also included representatives from the Girvan and Portpatrick Junction line, whose early vicissitudes were noticed above, which had running powers between Challoch Junction and Stranraer.

The Committee first confirmed the CR and Mr W.T. Wheatley as the operators of the Portpatrick and Wigtownshire lines respectively until other arrangements should be made to work them; it then debated what to do about existing plant and rolling stock. Many of the original locomotives, coaches and wagons which the Portpatrick Railway had originally acquired were still in use and were now CR property; to them the CR had added other vehicles. There was no problem there. Those used on the Whithorn line, however, belonged to the trustees of the late Thomas Wheatley and many still had his name painted on their sides. It was agreed to purchase these once W.T. Wheatley's period of management had ended and use them as long as they were serviceable. Sundry new appointments were made and certain officials confirmed in their posts.

The next meeting was 22nd October, when the main items on the agenda were who should work which lines and on what terms after 31st October, the choice of a new traffic superintendent and the termination of W.T. Wheatley's management of the line to Whithorn along with the purchase of its locomotives and rolling stock. As to the first matter, it was agreed that locomotive power should be provided by the CR and G&SWR, each agreeing to work certain trains each day and dividing as far as was feasible the mileage equally between them, and charges for these duties were also agreed. All four owning companies would supply vehicles; in practice the CR and G&SWR would provide the most. There were a few provisos regarding the use of shunting and ballast engines. The scheme was to operate for a trial period of one year. As to the traffic superintendency, it was thought desirable that the person chosen ought not to have been connected with any of the owning companies, so W. Cunning, who had worked on the Belfast and Northern Counties Railway in Ulster was appointed in place of Mr Grafton, who was given the position of traffic agent for the LNWR and CR at Stranraer until his retirement 9 years later. Thus the new broom began to sweep clean.

There remained the matter of purchasing the Wigtownshire plant, locomotives and rolling stock. This by now very rag, tag and bobtail assortment was valued at just under £5,200, which sum was offered to the trustees. W.T. Wheatley thought it was nothing like enough and managed to extract another £1,200; he himself received £166 13*s*. 4*d*. for his services from 6th August to the end of October, and was given the post of locomotive foreman at Stranraer. As will be seen later, he was to find other fields for his enterprise subsequently.

One immediate consequence of the sharing by the CR and G&SWR of the services for which the former had previously been responsible was that the shorter route between Dumfries and Carlisle was now followed by all through trains proceeding over the Joint Committee's lines, so that there was a saving of seven miles in distance and at least 20 minutes in time. From now onwards the Lockerbie–Dumfries line provided merely local and connecting services – though during the 'twenties and 'thirties of the following century a through coach from Edinburgh to Whithorn and back used this route for a while.

Over the next five or six years a good many improvements were made to the lines under the Joint Committee's control, some of which are considered in later chapters; others require mention here. There was the important question of water supplies. Local provision was often very inadequate. At Whithorn, for example, the storage tank had been built too low for water to have sufficient head to flush the station water-closets, and in 1886 it was in any case badly needing repair. At Newton Stewart the Burgh Council which provided the water had allowed the transmission pipes to get clogged. At Stranraer water was obtained from a well by means of a pumping engine, and the supply frequently ran dry. At Creetown it was obtained from a lady who owned a near-by estate and who was currently threatening to cut the supply off since she had not been paid for two years. There was also the matter of turntables. Whithorn did not have one, so half the mileage run by locomotives on that branch had to be tender-first unless tank engines could be found. Stranraer's turntable was only 40 ft long and would take nothing larger than a 2–4–0. Newton Stewart did not have one, neither did it have an engine shed. All these deficiencies were noted by the new officials, but lack of funds prevented much from being done to remedy them. Stranraer, however, was given a larger turntable in 1892, since engines from the Girvan as well as the Castle Douglas direction had to be turned there.

Soon after the Joint Committee's formation it began to press the Post Office for better terms for carrying the local mails, and in July 1886 a fresh agreement was concluded, by which about half as much again was to be paid annually; in return it was agreed that mail might be sent by any train and that the times of trains conveying letters would not be changed without Post Office consent. More important were developments in the mail service to Ulster. Discontent had been growing there for some years about the far from satisfactory service over the Holyhead–Kingstown route, and business men were complaining loudly. In 1891 the Joint Committee took up the cudgels on their behalf, saying it could provide a satisfactory supplementary service if adequately remunerated. Negotiations followed, and eventually a yearly subvention of £13,500 was agreed, to be divided between the Steamboat Company (which took the lion's share of £8,000), the Belfast and Northern Counties Railway, the Joint Committee and the G&SWR (which by now had acquired the line between Girvan and Challoch Junction).

Events on this latter line need a brief mention at this point since its trains used the Joint Committee's tracks for the last six miles into Stranraer and also ran trains down the Harbour branch. During 1885–1892 it was plagued by misfortune. At the time when the Joint Committee was formed the G&SWR was working the line from Girvan, retaining between 55 and 75 per cent of the gross receipts, according to whether they were high or low. In 1884 it had been able to pay a dividend of 1 per cent. The following year, however, the G&SWR informed the owners that it wished to terminate the agreement. It looked very much as if the intention was to force the closure of the line so that it could be bought for a song. The Girvan and Portpatrick Junction Directors then boldly determined to work the line themselves. They applied to Parliament for running powers over the existing G&SWR line from Girvan to Kilwinning and Kilmarnock (to both of which places the CR had already reached), sought permission to borrow £25,000 to enable it to work its own line, and asked for permission to sell its own line as a going concern if necessary.

Parliament granted all that was asked except the request for running powers, so a deal with the CR was not possible. Meanwhile the period of notice regarding the existing arrangement with the G&SWR had run out. The company kept a skeleton service running with hired vehicles, and also approached W.T. Wheatley, who had formerly managed the Wigtownshire Railway, and offered him the post of General Manager. Wheatley accepted, and used all his energy and influence to obtain second-hand locomotives and rolling stock, mostly from the North London Railway, ordered two 0–6–0s from a firm in Glasgow and showed that the little company meant to stand on its own feet and prosper.

The result was that offers to buy it began to come in. A London company bought the line in 1887 for £166,500 and altered its name to the Ayrshire and Wigtownshire Railway. The full train service now re-commenced, including boat trains connecting with the steamers to Larne. For three years the line succeeded in making a small profit, thanks largely to Wheatley's vigour, but the prospects were not bright; the line and vehicles began to deteriorate and some of the engines had to be taken out of service. Costs began to outpace revenue. Wheatley at last grew weary of the struggle and resigned, to buy and manage the George Hotel in Stranraer. The purchasers of the line began to wish they had not bought it. Then, somewhat unexpectedly, the G&SWR began to show an interest in it without first waiting for it to collapse. (This was probably because the CR was doing the same and rumours were circulating that it meant to build its own line to Girvan.) In 1892 the G&SWR bought the line outright and put it into a proper state of repair, and it became part of one of that company's main routes, with regular through boat trains from Glasgow.

As already mentioned, a steamer service had begun to operate daily between Stranraer and Larne in 1872. The Larne and Stranraer Steamboat Company had two paddle steamers on the service, the *Princess Louise* and the *Princess Beatrice*, which performed well enough, but the service did not attract enough custom to pay its way. One reason for the unprofitability was that the Post Office had so long delayed its decision to send mails by this route. By 1889 the *Princess Louise* needed to be replaced. This was made very obvious when an accident to the *Princess Beatrice* in January of that year made it necessary for her to be withdrawn for repairs. The *Princess Louise*, sailing in her place, simply could not keep the scheduled times. A new vessel was needed – but the company's accounts were in the red.

The problem was solved when the Joint Board obtained complete control of the steamship company's finances by buying out the other shareholders. Its own credit was then sufficient to enable the raising of the necessary loan, and a new ship was ordered, larger and more luxuriously appointed than the other two boats. The new *Princess Victoria* was to give satisfactory service (marred by a few accidents such as the breaking of paddle vanes in mid-passage) for many years, and was fast enough to provide two double crossings during the summer months. The *Princess Louise* was sold to David MacBrayne and Co. to run the Glasgow–Islay service, and survived on that duty until July 1902, when she struck the rocks of an island near Port Ellen in a thick fog and became a total wreck.

Princess Louise; the first steam packetboat to run regularly between Stranraer and Larne, seen here in 1905 berthed at Stranraer. *Stranraer Public Library*

Chapter Eight

Years of Success: 1893–1914

The two decades overlapping the end of the 19th and the beginning of the 20th century were a time of growing prosperity for the Portpatrick and Wigtownshire Joint Committee for many reasons – chiefly because of the greater demand upon its services for the transportation of goods. This was in large measure due to the more settled state of affairs in Ireland, and increasing prosperity there. It is true that there was growing political agitation there for Home Rule, out of which sprang a further demand for complete independence, but this was expressed in the Catholic south, not the Protestant north. Economically Ireland had greatly improved. Reforms by British Liberal administrations at Westminster had removed some of the grievances and given peasant farmers security of tenure, and measures taken by Conservative administrations to bring new industries to the poorer parts (colloquially known as 'killing Home Rule by kindness') had raised the standards of living in those areas.

Material prosperity partly resulted in an increasing surplus yield of agricultural produce and livestock, much of which, exported by way of Larne and Stranraer, brought revenue to the Galloway railways. Large quantities of fresh fish caught in the proximity of Stranraer were also carried by train eastwards towards the Borders and northern England, and some found their way further southwards in the special fish trains which were run after the turn of the century. The Galloway farms, particularly in the Rhinns and the Machars, marketed increasing amounts of dairy produce and beef cattle which were taken by rail to other parts of the United Kingdom. In the opposite direction large amounts of coal were brought from Tyneside collieries for use in locomotives and steamships as well as for domestic consumption.

At the same time there was a corresponding growth in passenger traffic. Tourism was beginning to flourish, and visitors began to seek out the scenic delights of the North of Ireland, such as the Antrim coast and the Giant's Causeway, travelling by way of Stranraer and Larne. The quieter attractions of the Galloway shores were a little later in exerting their influence, but here, too, hotels began to be built for tourists, and golf courses were established. Portpatrick in particular profited from its discovery as a holiday resort, and saw the building on its cliff-top of one of Scotland's most impressive hotels. Day trippers began to make their appearance and special excursions were advertised for their benefit. Stranraer became the objective of many Belfast people on a day's outing, once it had become possible to sail there from Larne in the morning and return in the afternoon.

The establishment of the supplementary mail service by the short sea route in 1891 brought the railway a regular income, as also to the steamship company, but problems accompanied it. Increasingly mails from the south were arriving late at Belfast, and by the end of 1899 the average delay had mounted to almost an hour. The loss of time was not the fault of the Scottish lines. Indeed, when trains were handed over late at Carlisle they often managed to recover a little of the lost time. The Post Office, however, threatened to withhold payment of their subsidy if the situation did not improve. The LNWR, the chief offender so far as late running went, blamed the Post Office staff where the mails were loaded at various stations on their line for their slowness in working. Another trouble was that the Belfast and Northern Counties authorities would not hold connecting trains to await the late arrival of steamers.

Letters travelling in the opposite direction were even more subject to delay, and by 1900 Belfast business men began to make representations to the Post Office. These gentlemen, as seen above, had previously been critical of the service provided by train and boat on the Holyhead–Kingstown route, and had asked for a service by way of Stranraer. Now that the service over the former route had greatly improved they hinted that, so far as mails to and from London were concerned, this was quite satisfactory, and that no objection would be raised by them to the discontinuance of the supplementary service. The Joint Committee, notified of this, realised that, whoever might be to blame for the situation in its own area, something had to be done about it, and especially about the southbound mails. Their lateness was being caused by the difficulty of combining the transport of perishables to the south with the conveyance of the mails. The former traffic had been increasing, and it was important to keep it, but equally important to retain the Post Office subsidy. Eventually the Joint Committee decided it must go to the expense of running a separate train to carry the perishables, and especially the fish traffic, timed to leave Stranraer after the mail train. It also decided to install automatic tablet catchers at most of the passing loops so that trains could exchange tablets at a somewhat higher speed. (In fact this, once done, made almost no difference as the trains habitually took the loops at a higher speed than was officially permitted, the fireman wearing a leather sleeve on his jacket to take the shock when he hooked the tablets up.)

It was not only from Belfast and London that protests came. The Joint Committee had to take into account locally-expressed wishes about the service it provided, and throughout its period of existence it had difficulty in reconciling them with the needs of the through traveller. The people of Portpatrick in particular were vociferous in their complaints. When the new hotel opened in 1904 one of the speakers at the opening ceremony laid into the railway, saying that if only a different pattern of train service were provided, more conveniently timed, many more people would use it. The Joint Committee's traffic manager, Mr Hutchinson, replied that the criticism, while understandable, was not really fair. The Joint Committee and the railway companies that worked its line had expended a great deal of money on it and received a small return. They would be only too glad to provide a better service, but the population of the district was sparse, and there were many connections with trains on other lines which had to be made. To alter the pattern of the existing service would only produce a worse one. His task was to get his priorities right, with the demands of the long distance traffic, which brought in the bulk of the receipts, necessarily being considered first.

At about this time a proposal was put forward to provide a steam rail-motor, of the type much favoured in Edwardian times, to work the service on the Portpatrick branch. It was estimated that it could be run at 7*d*. a mile, and if there were seats for 50 third class passengers, with smoking and non-

Royal Mail Paddle Steamer *Princess Victoria* seen here in 1904 gently approaching Stranraer Harbour.
Stranraer Public Library

Murray's Railway Guide passenger timetable for 1903.

Royal Mail and Shortest Sea Route between Great Britain and Ireland.—See Page 208.

GLASGOW, EDINBURGH, NEWTON-STEWART, WHITHORN, STRANRAER, PORTPATRICK, and BELFAST, via LARNE.

Leave	p.m	a.m	a.m	a.m	p.m	a.m	p.m	p.m
38 Glasgow { St. En.	11†10	..	..	..	..	1120	3 0	5 30
62 Glasgow { Cent..	..	..	..	..	..	12 0	..	4 35
62 Edinburgh PrinSt	a.m	..	..	..		12 5	2 15	4 25
40 Carlisle (S.W.)..	3 10	..	..	7 0		1p40	4 40	6 5
54 Dumfries	3 55	..	..	8 50		2 50	5 27	8 0
	a.m	..	..	a.m		p.m	p.m	p.m
Castle-Douglas lev.	4 26	..	..	9 45	Fridays only.	3 30	6 0	8 36
Crossmichael	..	..	..	9 54		3 37	..	8 44
Parton	..	..	..	10 4		3 44	..	8 50
New Galloway......	..	..	..	1012		3 50	6 12	8 57
Dromore............	..	..	..	1036		4 12	..	9 17
Creetown	..	..	..	1052		4 22	..	9 25
Palnure	..	..	..	1057		4 29	..	9 32
Newtn-Stewart arr.	5 8	.	.	11 4		4 35	6 43	9 38
Wigtown Br. Nwtn-Stewart lev	6 15	8 35	..	1120	2 30	5 0	8 0	..
Wigtown	6 31	8 59	..	1136	2 45	5 16	8 19	..
Kirkinner	6 37	9 8	..	1142	2 51	5 22	8 26	..
Whauphill	6 43	9 18	..	1148	2 57	5 28	8 32	..
Sorbie .... [stown)	6 49	9 28	..	1154	3 3	5 34	8 39	..
Millisle (for Garlie	6 55	9 37	..	12 0	3 8	5 40	8 45	..
Whithorn......"	7 5	9 47	.	1210	3 18	5 50	8 55	..
Nwton-Stewart lev.	5 11	8 45	Fridays only.	1111	3 35	4 41	6 46	9 46
Kirkcowan	.	8 56		1123	3 46	4 54	..	1012
Glenluce	..	9 10		1140	4 4	5 10	..	1025
Dunragit	..	9 16		1155	4 12	5 18	..	1031
Castle-Kennedy	..	9 23		12 2	4 21	5 26	..	1038
Stranraer Hrbr arr.	5 47	..		..	..	..	7 25	1046
Stranraer Stn { arr.	6 10	9 29		12 8	4 26	5 32	..	..
Stranraer Stn { lev.	7 0	..	1030	1215	..	5 42	.	..
Colfin	7 16	..	1040	1225	..	5 52	..	..
Portpatrick arr.	7 24	..	1050	1235	..	6 2	..	..
Stranraer Boat lev.	6*0	..	..	..	..	..	7 33	..
Larne ...... " arr.	7 46	..	..	..	..	..	9 15	..
Belfast ..Train arr.	8 35	..	..	..	..	..	10 0	..

Irish time.	a.m	a.m	a.m	a.m	p.m	p.m	p.m	..
Belfast Train lev.	..	..	9 5	..	..	..	5 30	..
Larne .....Boat "	..	..	9 45	..	..	..	6 20	..
Stranraer Hrbr ar.	..	..	1223	..	..	..	8 53	..
Portpatrick lev.	..	8 45	..	11 0	2 30	6 15	6 15	..
Colfin	..	9 0	..	1110	2 43	6 25	6 25	..
Stranraer Sn { arr.	..	9 10	..	1120	2 54	6 35	6 35	..
Stranraer Sn { lev.	7 35	9 30	..	Fridays only	3 40	6 50	8 15	..
Stranraer Hrbr lev.	..	..	1230		..	..	9 8	..
Castle-Kennedy ..	7 43	9 37	..		3 48	6 58	..	..
Dunragit	7 50	9 49	..		3 56	7 14	..	..
Glenluce	8 0	9 59	..		4 4	7 25	..	..
Kirkcowan	8 18	1017	..		4 24	7 42	..	..
Nwtn-Stewart arr.	8 28	1027	1 11		4 34	7 52	9 45	..
Wigtown Br. Whithorn ..lev.	7 25	9 25	..		3 30	..	6 50	..
Millisle (for Garl	7 34	9 37	..		3 39	..	6 58	..
Sorbie. [iestown)	7 40	9 45	..		3 45	..	7 3	..
Whauphill	7 47	9 52	..	..	3 52	..	7 9	..
Kirkinner	7 52	9 57	..	..	3 57	..	7 14	..
Wigtown	7 59	10 5	..	..	4 5	..	7 20	..
Nwtn-Stewart ar	8 14	1020	..	..	4 20	..	7 35	..
Nwtn-Stewart lev.	8 34	1037	1 15	..	4 41	..	9 48	..
Palnure	8 40	1044	..	..	4 49	..	..	..
Creetown	8 48	1053	..	..	4 59	..	..	..
Dromore	8 59	11 5	..	..	5 11	..	..	..
New Galloway	9 20	1126	1 53	..	5 33	..	..	..
Parton	9 27	1132	..	..	5 39	..	..	..
Crossmichael	9 37	1141	..	..	5 48	..	..	..
Castle-Douglas arr.	9 44	1148	2 8	..	5 55	..	1032	..
	a.m	p.m	..	..	p.m	..	p.m	..
54 Dumfries ..arr.	1026	1225	2 45	..	6 37	..	11 4	..
38 Carlisle (S.W.)"	1155	1 25	3 32	..	8 3	..	1147	..
64 Edinburgh, PSt"	..	4 5	6 0	..	1055	..	..	..
64 Glasgow { Cent"	..	4 0	6 8	..	11 5	..	..	..
40 Glasgow { St.E"	3 7	3 20	5 52	..	10 5	..	..	..

* 1st Class or Cabin Passengers can sleep on Boat the night before for 2/6 Cabin.
† Leaves 9-15 p.m on Sundays.
A Road Conveyance runs between Millisie and Garliestown in connection with all Trains.

Princess Maud was the first screw-driven vessel on the Stranraer to Larne route seen here in 1904 on Loch Ryan.
Stranraer Public Library

smoking sections and space for luggage, and if the locomotive were strong enough to haul up to two trailer-coaches, this would be a good solution to the branch's local needs. It would have been an interesting experiment; it would certainly have needed a tough little engine to take a coach and two trailers over those fearsome gradients. However the two chief holding companies, the LNWR and the MR, did not approve of the idea, so nothing was done about it.

It has already been mentioned that a line from Stranraer southwards along the Rhinns, having a similar object to that of the Wigtownshire Railway, was under discussion during 1877–1882, though it never came to anything. A similar scheme was mooted in 1898. It was suggested that a light railway be built, extending to Drummore in the south but dividing at its northern end, one branch linking with the Portpatrick branch between Stranraer and Colfin, the other making a junction with the Joint Committee's line near Dunragit. It was thought that it could be built for a little under £100,000. The scheme's chief promoter was a prominent Glasgow business man, Mr H. Mayberry, a native of Drummore, who found a few supporters locally, notably W.T. Wheatley, who was now an hotel proprietor at Stranraer.

An application was made for an order under the Light Railways Act of 1896, and in May 1899 two Light Railway Commissioners came to Stranraer to hear evidence. It then appeared that local opinion was mostly unfavourable, and the Board of the Joint Committee said they would not wish to work the line if it were built. Fears were expressed that it would damage the farms which supplied dairy products to the creameries. It was emphasised that coach services already operated catering for passengers and goods. Local farmers had become accustomed to carting their produce to Stranraer, and cattle were driven there on the hoof. The Commissioners therefore refused to recommend an order, and the scheme died. One has to agree with the dissenting majority, for a reason that could then not be foreseen. The age of the internal combustion vehicle was at hand, and the railway, had it been built, would soon have become unprofitable.

It was in 1907 that the first petrol-driven omnibus made its appearance in the Rhinns, and a very strange vehicle it was. There had been a horse-drawn mail coach service between Stranraer and Drummore since 1892, two coaches travelling each way daily and taking between 2½ and 2¾ hours to cover the 18 miles. In 1905 the operation of the mail service came under discussion locally, in the context of its operation by a motor vehicle. The first intention was for a limited company to be formed with a capital of £3,000 made up of £1 shares, but enthusiasm for such a project was lacking, and when the Portpatrick and Wigtownshire Joint Committee offered to run such a service with steam-propelled cars there was general approval. The County Council said it would widen the road, and orders were placed for two vehicles from the Kilmarnock works of the G&SWR. They were completed by April 1907 and taken to Stranraer to be stored there till the time came for beginning the service. Each was able to hold 30 passengers in a large compartment behind the driver's seat.

A contemporary photograph shows one, with bowler-hatted gentlemen gazing at the camera through large glass windows and a conductor in uniform looking out from the rear platform. Each vehicle was 25 ft long and 7½ ft wide; the tyres were of hard rubber; the driver, with two passengers beside him, held a slightly tilted steering wheel behind the square bonnet which enclosed the engine. Petrol was used as a fuel, in a decidedly unorthodox way. A burner, fired by petrol, heated a tubular coil, boiling the water within it and converting it into superheated steam, which after operating the pistons in the cylinders, was then condensed back to water; this was then fed back into the tubular boiler by an injector.

The new service began on 1st May, 1907 with a twice-each-way-daily service of runs between Stranraer and Drummore; the journey took 1½ hours. During July and August a mid-day service was also added. Reporting on the first month's operation, Hutchinson said that the public had given good support and that the competing horse-drawn omnibuses had been obliged to lower their fares. However, he also mentioned the frequency of breakdowns, and this was perhaps why, after the first half-year, the service had made a slight loss. This tendency continued, and the following year the Joint Committee ordered the ending of the service and the sale of the two vehicles. No one appeared to want to buy them, and eventually they were disposed of to a London firm for £100 each, one-tenth of what they had cost to build. The whole enterprise had proved an expensive failure. However, a few years later a Stranraer garage put on a service with motor omnibuses using petrol in the more conventional way, and gained the contract for carrying the mails.

The steamboat service between Stranraer and Larne during this period saw the introduction of new vessels to replace the older ones. The first was *Princess May*, a sister vessel to the *Princess Victoria* but slightly larger and one knot faster. On her maiden voyage she made the return crossing in record time, taking just under two hours quayside to quayside; conditions must have been very favourable to make this possible with a paddle steamer. She became a popular boat, but had her share of teething troubles, particularly with cracked paddle vanes; the *Princess Victoria* suffered similarly.

Despite practising stringent economies, the steamboat company was not paying its way, the main reason being the variation in patronage over the years; boats that ran nearly full in the summer might carry only a handful of people during the winter. From 1898 onwards attempts were made to attract custom by reducing the single fare between London and Belfast by a few shillings and by offering more substantial reductions with fortnightly return tickets, but these do not seem to have had much effect and were withdrawn in 1902. A later and more successful device to attract travellers was the installation of sleeping berths, both ships having a number of cabins so fitted. Though they did not sail by night, the evening boat from Larne reached Stranraer at a late hour, and the corresponding sailing left early to return, so anyone who wished to avoid a long night journey by rail could sleep on board while the ship was in port at Stranraer, either before the crossing (if coming from London) or after (if coming from Northern Ireland), making the rail journey by day. The scheme proved moderately attractive to passengers; it made a small profit during the year of its inception, 1898–1899, and continued to be reasonably well patronised in subsequent years.

Another money-raiser used from 1895 onwards was to employ whichever boat was not being used on the scheduled crossing to run special excursions, and during the holiday seasons these attracted much custom.

While *Princess Victoria* and *Princess May* were fairly fast boats the reserve vessel, *Princess Beatrice*, was much slower and incapable of keeping to the advertised schedules. As the other two ships grew older the company became concerned about their possible deterioration and commissioned a report on them. The result was a decision that both should be re-boilered in 1904, and that a third ship should be obtained, with screw propulsion to avoid the breakdowns which were occurring with the paddle steamers, and to enable it to draw up more closely alongside the quays than was possible with side paddle covers. In February a new 3-screw steamer, the *Princess Maud*, named after Edward VII's youngest daughter, 300 ft in length, turbine-driven and capable of over 20 knots, was put into service. She proved a success, making the passage on one occasion in 8 minutes under the 2 hours. However, she was heavier on coal than the other two ships. From 1905 onwards she was the regular passage boat. She was much liked by passengers, the day service being especially well patronised, particularly during 1905 and 1906. Unfortunately the company's annual losses continued, though they showed a tendency to decrease. Meanwhile the older ships were disposed of one by one. In 1904 the *Princess Beatrice* was sold for scrap, and 6 years later the *Princess Victoria* followed her. The latter's disappearance was much regretted locally; there was a feeling that if she had been re-boilered she would have been as good as ever.

Down now to two boats, the company chartered a third, the *Duchess of Argyll*, from the Caledonian Steam Packet Company as a reserve ship, and ordered another, a sister vessel to the *Princess Maud*. It was ready in April 1912, and was named the *Princess Victoria* after its predecessor. The familiar pattern, of an early morning and late evening mail service all the year round, with return daytime sailings during the summer months, continued, and in 1914 yet a third daily return crossing was instituted between Whitsuntide and the end of August. By now the motor car was a more familiar sight on British roads, and from 1907 the company undertook to convey cars and motor cycles on condition that their fuel tanks were empty. It was therefore necessary to provide petrol storage tanks on the piers at both ends of the crossing. In 1914 it cost £1 15*s*. 0*d*. to take a car across – 14 times as much as the steerage fare. At the beginning of the same summer attractive special excursions were advertised along the coasts of County Down and County Antrim. Both the Railway and Steamship companies were now at the zenith of their fortunes. No one foresaw that the black cloud which hung over the Balkans at the height of the season presaged a disastrous war which would change European society for ever.

Chapter Nine

The First World War and after: 1914–1922

An immediate consequence of the outbreak of World War I was that all the country's railways, large and small, were temporarily nationalised, the Government taking over their management and control in the interests of military necessity, and placing them under the Railway Executive Committee, a body composed of senior officials from a number of the larger lines. Each company was guaranteed its 1913 dividends; during the course of the war the total amount paid out was £95 million.

The effect on the Portpatrick and Wigtownshire Joint Committee was at first small. A Saturdays-only service between Dumfries and Stranraer was cancelled in both directions, and all steamer services across the North Channel except the mail services were stopped. This, however, did little more than anticipate the winter timetable which came into effect on 1st October. At the end of 1914 the Joint Committee even published its annual Tourist Guide, as if nothing had happened. This was in fact the last such guide to be put out under the Committee's auspices.

The War was not at first expected to last very long. However, once the German invasion of France had been brought to a halt in September 1914 and the trench lines had been established, it was realised that the struggle would be lengthy and more drastic changes came about. To consider first the steamship service: the number of boats available for making the passage to Ulster was much reduced. The *Princess Victoria* was the first to be taken away; she was requisitioned in October 1914, transferred to Southampton and used there as a troop transport. The next vessel to disappear was the *Princess May*. It had in any case been the Joint Committee's intention to dispose of her, and she was sold for scrap and made what was expected to be her last voyage on 10th December, 1914 to South Wales. However, the Admiralty interposed and bought her and used her on special duties. At the end of the War she was laid up for a while before being broken up at Liverpool. Only the *Princess Maud* was left. With the aid of a steamer loaned by the LNWR as a stand-by, the *Galtee More*, the Larne–Stranraer service was maintained. The year 1915 passed without incident, except that right at the very end raids by German Zeppelins commenced, and with exaggerated caution trains and boats were halted for long periods, even when the potential bomb-dropper was far away on the East Coast.

The Stranraer–Larne link was especially important during the War, as it was a route less open to enemy harassment by submarine activity than the other Irish Sea crossings. The political state of affairs in Ireland necessitated a reliable connection. So both ports attracted more traffic in passengers and goods than otherwise would have been the case. This showed up in the inadequacies of the unloading arrangements at Stranraer. The port had two cranes, one built in 1878, the other in 1896, but neither operated rapidly enough to enable a full cargo to be put into the ship or be discharged from it during the eight hours when it was moored to the quay between sailings. The consequence was that goods loaded at Larne frequently could not be taken off and had to make a number of journeys to and fro before being at last removed. Before the War a new crane could have been obtained fairly quickly, but all the country's engineering factories were now given over to war work, and it was not until June 1918, when the end of the war was almost in sight, that an adequate crane was provided.

During the War years Army and Navy personnel were stationed at various places in Galloway, and two naval vessels were based in Loch Ryan. More passengers were therefore carried on the railway, as soldiers and sailors travelled home on leave, and some unusual goods were also transported. Large quantities of hay from farms in Wigtownshire were loaded into wagons and hauled away to be used as fodder for Army horses. Timber, which had formerly been obtained from Scandinavia for use as pit props, could no longer be brought in from that quarter, so local Kirkcudbrightshire forests were drawn upon instead. The Joint Committee's staffing was also affected, women taking the places of younger men who volunteered or were called up for military service. As was to be expected, holiday-making in the area declined. Fishing parties no longer came to fish the Galloway streams, and grouse shooting became less of a business now that there were other targets to aim at in France and elsewhere. During 1916 an unusual activity was established – the gathering of sphagnum moss for use in medical dressings. This centred on Loch Skerrow passing place, where persons in the Joint Committee's employ went in their spare time to rake away the heavy wet moss from the patches where it grew on the moors, and were given free meals in return. The moss was dried, put into sacks and despatched to hospitals.

At the end of 1916 the Government requisitioned 300 railway locomotives to be transported to France for use in military transport, and this affected railways all over the country. In Galloway it became necessary to run fewer trains and to lessen their loadings. Only one first class sleeping car now ran between London and Stranraer, in which all the berths but two were reserved for cross-channel passengers. Rail fares were put up at the same time in an attempt to deter people from travelling, but as it happened this did not have much effect. During 1917 and 1918 the risk from submarine attack was at its height, and Allied vessels were being sunk at sight in the Irish Sea. The *Princess Maud* was accordingly fitted with wireless transmission apparatus, and when sailing was escorted by a naval vessel and instructed to follow a zig-zag course, which of course added greatly to the passage time. She was also dazzle-painted to make her a more difficult target. In the event no attempt was ever made to sink her, but it was not till four days before the War ended that she was again permitted to sail on her normal course. She had survived some heavy use; her carpets, between six and eight years old, were in places worn right through.

The end of the War found the Portpatrick and Wigtownshire lines badly in need of attention and repair. Normal maintenance had not been possible, and a number of derailments occurred because of the poor state of the track. The immediate post-war period was a difficult time, since the future of the whole railway system was under consideration and no individual company could know what its fate would be. A Select Committee on Transport had been set up by Parliament in 1918, which concluded that 'unification of the railway system is desirable under certain safeguards, whether the ownership be in public or private hands'. It did not look as though the *status quo* was going to be perpetuated. Pro- and anti-nationalisation interests argued it out behind the scenes, and at the end of 1920 a White Paper showed that the Government now favoured a grouping of the railways under Boards of Management each composed of representatives of shareholders and of employees. This scheme was disliked both by the Railway Companies and their shareholders, and by the railway trade unions and the Government was forced to think again. The scheme finally adopted largely stemmed from the powerful mind of Sir Eric Geddes, who influenced the drafting of the Bill which eventually proposed the division of the railways in Britain into four large groups. The Bill became an Act in 1921, the day appointed for its implementation being 1st January 1923. Now everyone knew what to expect. The Portpatrick and Wigtownshire Joint Committee was swallowed up in the new London Midland and Scottish system as from that date.

During the four post-War years before the groupings a little, though not much, was done to improve the line. In 1913 the decision had been taken in principle to re-align all the crossing loops between Castle Douglas and Stranraer to allow faster running for non-stopping trains. The outbreak of the War put an end to such projects, and in 1916 a less thoroughgoing alternative was fixed upon; long section tablet working would be established so that there would be fewer places for token exchange and trains could run faster between them, always taking the straight line at the crossing loops which had been switched out. This would result in a quickening of schedules, and also in a saving of wages, since so many signalmen would not need to be on duty at any one time. The existing automatic exchange system would be discarded.

Between Stranraer and Castle Douglas there would be only three changing places when long section working was in operation – at Challoch Junction, Newton Stewart and Loch Skerrow. Arrangements were to be made for switching in the other loops in an emergency, loud-sounding electric bells being installed in the stationmasters' houses, operable from any one of the signal boxes, so that warning could be given of the need to revert to short section working if, for example, an engine failure should occur. Such a system, while permitting faster running, would also limit the number of trains which could be run; however, with only four or five through passenger trains each way a day the longer sections would have sufficed during the night hours. This system was mooted, but not actually brought into use until after the groupings, as described in the following chapter.

The final years of the Portpatrick and Wigtownshire's independent operation saw a revival of the scheme to build a light railway from Stranraer to Drummore, the suggestion this time not arising locally but coming from a Government Committee established to suggest improvements in rural communications in Scotland. A public meeting was called in support of such a line, which met at Stranraer on 1st July, 1920. The plan was for it to link with the line between Challoch Junction and Stranraer at Dunragit, and run south-westwards across the head of Luce Bay and thence south along the eastern coast of the Rhinns. It was understood that a Government subsidy would be forthcoming. This assurance having been given, the meeting approved the scheme. But in the end, like the two previous schemes, it came to nothing. A short-lived nation-wide boom following the end of the War had been succeeded by a slump, and the Government was not disposed to keep its promise.

Meanwhile, the steamship service came slowly back to the 1914 pattern of operations. The *Princess Victoria* was released by the Admiralty at the end of 1919 and, after being reconditioned, was put into service again between Stranraer and Larne in the spring of 1920. The extra daily services, however, were not yet restored. Both the *Princess Victoria* and the *Princess Maud* were now fitted with wireless receiving and transmitting sets. They were kept busy during the post-War boom, which lasted till the end of 1920, though a miners' strike at one point cut off their supply of coal and forced a cessation of the service for a week.

Thus, for the Portpatrick and Wigtownshire Joint Committee, with its strong local links and centre of management at Stranraer the sands finally ran out. At the beginning of 1923 it was absorbed into the London, Midland and Scottish Railway, and was operated from outside its own area. New locomotives, hailing from other parts of the LMS system, appeared on its metals. Unfamiliar coaches, cascaded from other lines on that system, replaced the former CR and G&SWR ones. One by one the Joint Committee's staff disappeared, by retirement or transfer. Personal relationships were replaced by more impersonal ones and the feeling of mutuality vanished.

A goods train, bound for Stranraer, hauled by a Stirling 4–4–0 seen here in 1930 crossing the massive structure of Glenluce viaduct. *Stranraer Public Library*

Chapter Ten

Under LMS Management

Although the groupings in 1923 made the former Portpatrick and Wigtownshire lines part of an enormous system which extended from London and Bristol in the south to the extreme north of Scotland, with consequent removal of local management, this did not result in a worsening of facilities. Things were bound to improve as wartime conditions gave place to those of peacetime, and repairs and maintenance could at last be undertaken. However, the growing realisation that the separate private companies were soon going to lose their identities had not been conducive to vigorous attempts at recovery. The state of the Portpatrick and Wigtownshire lines' permanent way remained unimproved during the four years which followed the Armistice, and the derailment of a train on the Cree viaduct at the end of December 1922 highlighted its condition. The Inspecting Officer's findings at the end of the accident enquiry included a condition that a speed restriction of 45 mph should apply over the whole line until the track could be brought up to a satisfactory standard. This meant, for a while, somewhat lengthened schedules.

After the LMS had taken over the line the work of relaying it was begun, and by 1926 was completed. Nevertheless caution was still needed. No heavy locomotives were allowed over it for some time. Meanwhile, in 1924 the long-intended introduction of long-distance tablet working for night trains between Castle Douglas and Stranraer was begun, with some changes in details from the scheme previously approved. During the day the existing signal boxes and crossing loops were still used, but during the night hours those at Castle Kennedy, Glenluce, Kirkcowan, Palnure, Creetown, Gatehouse of Fleet, New Galloway and Crossmichael were switched out, only the ones at Dunragit, Newton Stewart and Loch Skerrow remaining operative. At the same time the Manson apparatus at linesides and on locomotives was also dismantled. Now that the line was under LMS ownership the number of locomotives which might be assigned to work on this section was much greater, and the fitting of the necessary gear to all those which might possibly be required to work there would have been an expensive proposition. Hand exchange was reverted to, which of course meant that non-stopping trains had to reduce their speed further in order to set down and take up tablets. However, in practice it was only a matter of a few miles an hour, and was probably not a great cause of delay.

Since its inception the Portpatrick Railway had never had any Sunday services. Local opinion, vigorously supported by church ministers, was against it. However, the War had relaxed many peoples' convictions in regard to Sabbath observance (one could not, after all, refuse to fight or do necessary war work on one day of the week) and there were no protests made when the occasional running of Sunday excursion trains began. One innovation in 1927 was the running of special trains for day trippers during summer Sundays between Glasgow and Portpatrick. The first of these was somewhat of an event. On 19th June a 12-coach train of corridor stock, full of Glaswegians, left St Enoch station for Portpatrick, hauled by two rather small Manson 4–4–0 engines. The severe gradients tried the latter greatly between Ayr and Stranraer, but the real test came on the extension to Portpatrick, which had previously been accustomed to supporting one or two coaches and a tank engine on weekdays and nothing at all on the Sabbath. Arriving at Portpatrick an hour and 20 minutes late, the train disgorged its army of excursionists (who presumably, if at the rear, had to walk through the vestibules in the front coaches before they could alight) and was then propelled back to Stranraer, where its coaches were cleaned and serviced and the locomotives turned and re-stocked with water and coal; the latter then pushed the train back to Portpatrick to collect its human cargo. It was ¾ hr late leaving there and 2½ hours late arriving at St Enoch. The experiment was repeated a week later with rather larger engines and a slightly lighter train, and on this occasion time was kept. Such excursions were run many times subsequently, though Portpatrick was not the only destination. One wonders what the local Kirk ministers felt about this invasion of their Sabbath calm.

The two decades between World War I and II saw a gradual increase in traffic over the Stranraer–Larne route and the provision one after another of new vessels to ply upon it. The pre-war *Princess Maud* and *Princess Victoria* maintained the service for some years, but both were now growing old, so in 1931 the former was sold for scrap after a replacement ship had been built. This new vessel, the *Princess Margaret*, named after George V's young grand-daughter, was a turbine-driven steamship with twin screws, over 300 ft in length and displacing 2,552 tons. For a short while she ran in conjunction with the *Princess Victoria*, but the latter was found increasingly inadequate, having no sleeping berth provision, so she too went to be broken up in 1934. A new *Princess Maud*, similar to the *Princess Margaret*, took her place.

One feature of the 1930s was the gradual increase in car-ownership, though this was not yet a general feature of society but confined to the better-off. Naturally many car-owners liked to use their vehicles for holiday touring, and the scenic delights of Northern Ireland were an attraction, but the existing ships were not adapted for the easy loading and unloading of automobiles, and it was necessary to carry them on deck and load and unload them by crane – a slow business and permitting only a few to be shipped on any one sailing. By 1938 the cross-channel traffic had increased so greatly, and the demand for car-shipment was so considerable, that a new vessel able to load cars through stern doors was required. In 1938 it was ordered, and completed by July 1939. It was about the same size and tonnage as the *Princess Maud* and fitted up to accommodate 88 cars. It had no sleeping accommodation, so was used on the daylight service, the *Princess Maud* being used to connect with the night trains.

However, the new *Princess Victoria*, the third to carry that name, was to have a short life and an even shorter one on the service for which it had been designed. Two months after it began to operate World War II broke out, and both she and the *Princess Maud* were commandeered by the Government. The latter was used as a troop carrier between Southampton and France, and while so serving was damaged by an enemy shell which killed several of her crew, though she herself survived. *Princess Victoria* was converted as a minelayer, and in that capacity served off the East Coast until being struck by an enemy mine off the mouth of the Humber, when she sank with all hands. Meanwhile the *Princess Margaret* had been brought back from the Heysham–Belfast run, where she had been helping out during the summer season, and kept the Stranraer–Larne crossing going throughout the War years, being used only in connection with the night trains; there were no scheduled daytime services until hostilities had ended.

So far as the former Portpatrick and Wigtownshire lines were concerned the War altered everything. There was no temporary pretence, as there had been in 1914, of 'business as usual'. The holiday traffic vanished. Indeed, ordinary travel was officially discouraged; posters on station notice boards all over the country enquired, 'Is your journey really necessary?' Traffic from Stranraer to and from both Glasgow and Carlisle now largely consisted of the movement of troops.

When the War began the Government of Southern Ireland declared that it was firmly intending to remain neutral; this was interpreted by the British Government as the expression of an unfriendly attitude and, rightly or wrongly, it was suspected that a German attack might be mounted from that quarter. A whole division of Welsh troops was therefore stationed in Ulster as a safeguard. The transfer was made in April 1940, 36 trains being required to perform the operation, and from then on there was much coming and going of members of the forces serving in Ulster, in special leave trains, two running each day between Stranraer and Cardiff or London. Servicing these enormous 16-coach trains was beyond the resources of Stranraer, so they had to be taken to Ayr for this purpose and then brought back the same day – a gigantic job for the two ancient ex-Highland Railways 4–6–0s which had been due to go to the scrap heap before being given a stay of execution. At slow speeds, and working in tandem, they hauled the enormous loads of empty coaches over the double journey, 60 miles each way, to Ayr and back. During the War siding space was at a premium in Stranraer, and the continuation of the line towards Portpatrick was used as an additional siding in which trains could be stored, one in line with another.

The movement of troops to and from Ulster was accompanied by the movement of Army vehicles, including tanks. There were no facilities on the quay at Stranraer to deal with such weighty objects, and the available ships could not readily carry them even if they could have been loaded on. However, suitable vessels did exist – the three ferry ships which during peace time had been used on the Dover–Dunkerque crossing to take through wagons and Wagons-Lits vehicles to and from the Continent. These, the *Hampton Ferry*, *Shepperton Ferry* and *Twickenham Ferry*, were brought round the coast to Stranraer, and for four years conveyed military vehicles across the North Channel, using the slipways that had been prepared for the carriage of cars when the *Princess Victoria* had been put into service.

During the War other parts of Galloway became temporary Army, Navy or Air Force centres which contributed to the total of service men regularly travelling on leave by rail. There was a seaplane base on the west side of Loch Ryan; there were two airfields, one on the east side of the same loch and one near Wigtown in the Machars; there was an artillery range south of

Whithorn near Burrow Head; there were also two war factories temporarily set up near Newton Stewart. So the former Wigtownshire Railway, too, had its share of war traffic. The demand for the area's agricultural produce was even greater than before now that food from overseas could only be obtained through risk of submarine attack.

Most important of all the temporary wartime changes in this region was the construction of an emergency port at Cairnryan, a few miles from Stranraer up the eastern side of Loch Ryan. This was intended to function as a substitute for Glasgow or Liverpool in case either was put out of use through enemy bombing. A line of railway was built to link this port with the former Portpatrick Railway a mile east of Stranraer. The whole installation was completed by July 1943 and War Department locomotives were brought in to work the line. Cairnryan Port had deep water wharves with numerous cranes and extensive sidings. Little use was in fact made of it. When the War had ended it was used as a shipbreaking centre for a while, and also for getting rid of a large store of gas shells for which fortunately no use had ever been found; these were loaded into old ships which were then taken out and scuttled in mid-Atlantic. Eventually the War Department gave up the port into civilian hands, and the railway was dismantled; the wharves remained and were later used for ferries conveying cars and lorries to and from Larne, supplementing the service provided from Stranraer.

After the end of World War I the former railway companies had had a 4-year period in which to get used to impending change. The LMS had only 2½ years. The Labour Party, which won the 1945 General Election, had included in its manifesto an undertaking to bring the whole railway system under public ownership, so from the summer of 1945 onwards the writing was clearly on the wall. There was little time for changes or developments before Vesting Day on 1st January, 1948, the 25th anniversary of the formation of the LMS and also the date when it ceased to be. The most important of the events of 1945–1947 was the building of a new stern-loading car ferry for the channel crossing, an oil-driven vessel constructed in Glasgow by Denny and Co., and named like its three precursors *Princess Victoria*. It worked the daylight service in both directions during the summer from 1947 onwards, and in the winter months took over the early morning and late evening run. It was the finest and most up-to-date vessel on any of the Irish crossings; tragically, its career was cut short when it foundered in a storm during January 1953. An account of this event is given in Appendix 6.

Chapter Eleven

Under British Railways: Decline and Closure

The last 17 years of Galloway's railways after nationalisation present a gloomy spectacle, when a final battle was fought with the invincible motor vehicle, which turned out to be a series of retreats. After World War I motor transport on British roads had indeed increased, but not so seriously as to threaten the viability of the railway system. Once World War II, in winning which the railways had played a vital part, was ended, and a measure of national prosperity had returned and average personal incomes had begun to rise, the lure of the private car became irresistible. Car ownership spread down through society until it became, for most families, almost regarded as a necessity next only to the acquisition of a house. In part the car was a convenience, in part a status symbol. Once such an expensive object had been bought, it had to be used, and if one were making a journey it cost no more to use it than to use public transport, and if one carried passengers it cost far less than buying a number of rail tickets. Thus arose the situation in which many children, though travelling on holiday frequently, had never seen the inside of a railway carriage. It also became increasingly more economical to use road transport for the carriage of goods, especially when the distance was not great and delivery could be made to the door. Rail remained the preferred option for goods in bulk, such as coal, but the old method of transporting miscellaneous items in wagons or vans, which took their time between various shuntings and marshallings, and might spend days on the journey, passed away.

British Railways, when incorporated by Act of Parliament, was given the obligation of trying to make ends meet overall. Given the traditional methods of working and signalling, which had been developed in Victorian times, when labour was cheap and technical skills were in their infancy, this was impossible for local branch lines. In country places, if one did not have a car, the bus service which could pick you up near your starting point, or put you down closer to where you wished to be, was more convenient than a railway where the station might be more than a mile away. If one merely wished to be locally mobile, too, one did not have to buy a car; there were motor-cycles, scooters and mopeds to be had. So people had less recourse to the railway, and this was certainly the case in Galloway.

The first branch line to go in Galloway was the line from Stranraer to Portpatrick, between which places buses were offering a service which was as fast and more convenient. It closed on 6th February, 1950, and soon afterwards the track was taken up between Portpatrick and the summit station at Colfin; from the latter to Stranraer itself it was allowed to remain for the time being to serve the creamery that had been established there.

The Whithorn branch was the next to feel the axe. After 75 years of operation the last passenger train ran from Newton Stewart to Whithorn and back on 25th September, 1950. A local resident who travelled on it described, in a privately-printed book, the mixed feelings of most of the passengers, and the unmixed feelings of two elderly sisters who lived at Whauphill, which no buses served. They had long been in the habit of taking the Saturday afternoon train to Newton Stewart, doing their week's

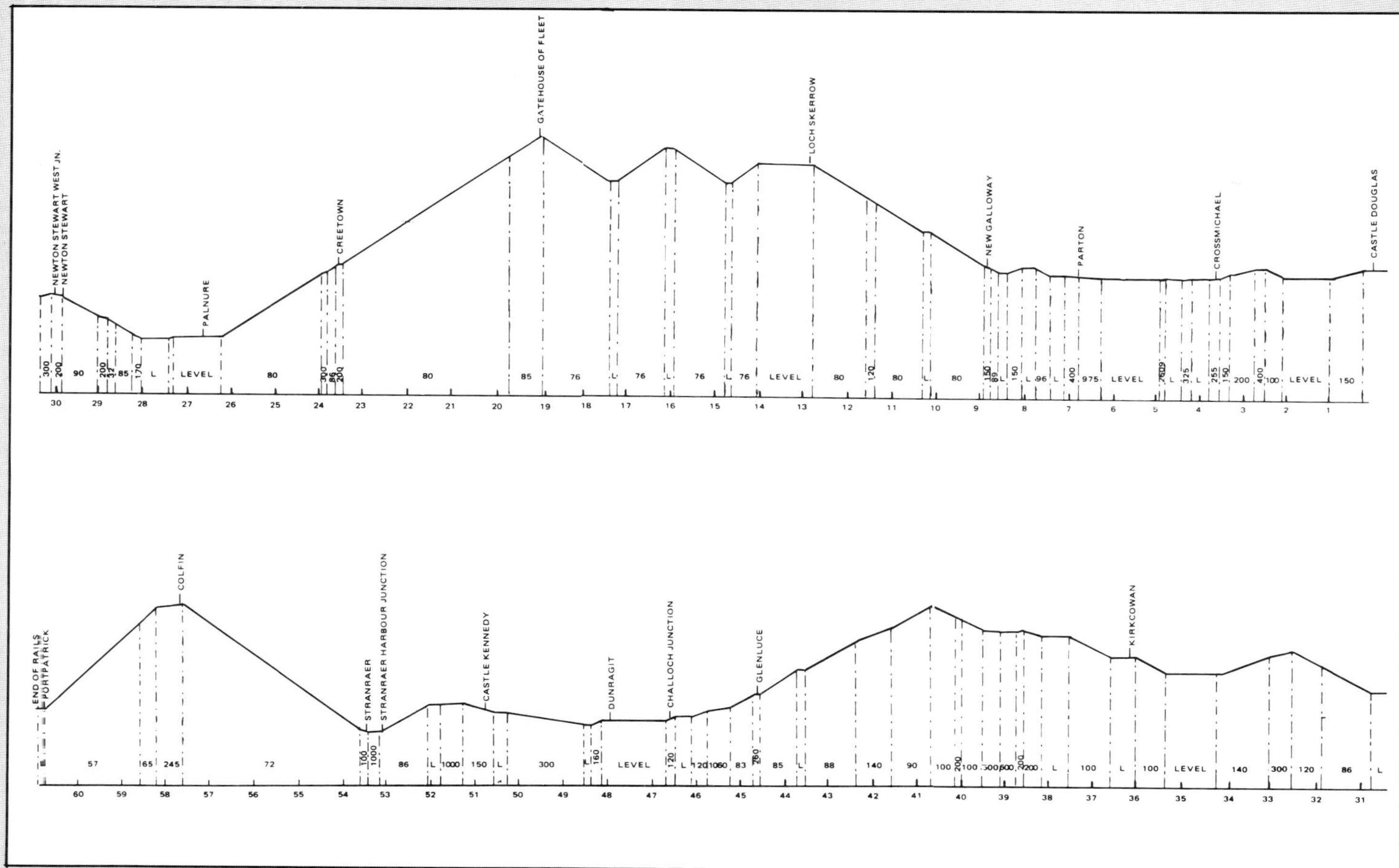

GRADIENT PROFILE FOR THE PORTPATRICK RAILWAY.

shopping there, having tea together in a café and then going to the first house in the cinema to watch a picture before returning on the last train at 8 pm. They were angry at the change which had upset the regular pattern of their lives. One wonders how they subsequently coped. No doubt a TV set eventually took the place of the movies, and one can make one's own tea. Shopping became less necessary when vans came to deliver groceries at your door. A railway cannot be kept operating at a loss just to please a few old ladies, and the pair must often have looked at the goods trains which continued to pass through Whauphill twice each way a day, and perhaps wished they could have thumbed a lift. The line was kept running for freight alone for a further 14 years. When no passengers were carried economies could be made; signal boxes were replaced by ground frames worked by the footplate staff, and the number of block sections was reduced to two.

As to the main line, there seemed at first no cause for apprehension; it was, after all, one of the main routes from England to Northern Ireland, and the preferred one for those who wanted a short sea crossing. Two of the stations were nevertheless deemed redundant. Gatehouse of Fleet, six miles from the burgh it claimed to serve, at the end of a long narrow road, seemed an obvious case for closure. Whatever might have been the case in the 19th century, it seemed un-necessary now, when people who wished to take a train eastwards could just as conveniently get the bus to Kirkcudbright and pick it up there, and those going west could do the same to Newton Stewart. On 5th December, 1950 the station was closed to all traffic – but, then, soon afterwards re-opened because of the hardships felt by the few people who still lived near it and for whom it was the only means of transport. One up and two down trains therefore continued to call there. Palnure was also closed on 7th May, 1951, and there was no call for it to re-open; adequate bus services existed. The other stations on the line remained open, though some must have had very little custom.

On 1st April, 1959 the line from Stranraer to Colfin was closed to all traffic, the creamery now despatching its products by road. Along with the closure and removal of the track went the demolition of Piltanton viaduct, which was blown up.

Unlike the branch lines, the steamer service was prospering, since it not only received passengers from the railway, both from the Glasgow and Carlisle directions, but also transported cars which did not travel on the railway and so brought no revenue to it. The *Princess Victoria* catered for this traffic. Early in 1952 the *Princess Margaret* was reconditioned and given oil-driven engines; it was then proposed to use her for taking the mails directly to Belfast from Stranraer and keeping the *Princess Victoria* for the daily run to Larne, which was to be an all-the-year-round service. The suggestion aroused strong opposition on both sides of the water and was dropped. Early the following year disaster struck the shipping company, and the whole community at Stranraer, when the *Princess Victoria* was lost with all hands and most of her passengers in a violent storm in the North Channel on 31st January, 1951. (See Appendix 6 for a fuller description of this, the worst ferry disaster around British shores until the overturning of the *Herald of Free Enterprise* in 1988.)

After this tragedy the question of the continuance of the Stranraer–Larne sailings became acute. It was a profitable service, so long as the boat in use could carry motor vehicles in sufficient numbers so that there were no long waiting lists and one could more or less drive on and drive off. For some years the matter was in doubt. The *Hampton Ferry* was brought back from the south and kept the service going, but she was now an ageing vessel and had not been built for cars but rail vehicles. The *Princess Margaret* kept the mail service running meanwhile. In 1960 the decision was made to acquire a new boat, which was ordered from Denny and Co. in Glasgow. However, it was four years in building, and when it was finally delivered the former steamship company had been wound up and replaced by a new one, the Caledonian Steam Packet Company (Irish Services) Ltd. This firm took delivery of the *Caledonian Princess* in December 1961. She was the largest ship ever to have been put on the Stranraer–Larne run, being 353 ft long and displacing 3,630 tons; she could carry 1,400 passengers and over 100 cars, and was well supplied with sleeping berths. The vessel began a service of two journeys each way daily throughout the year, with one on Sundays during the summer season. From that time onwards the future of this particular crossing was assured, though the writing was on the wall for the old Portpatrick line which had for so long supplied it with non-motorist passengers from the east and south.

Soon afterwards British Railways was given a new Chairman from the world of business, Dr Richard Beeching, whom many even now love to hate. Under his advice and influence a report was produced on *The Re-shaping of British Railways*. The general intention was to cut away services which promised no profitability and to concentrate on those which could be profitably developed. The railways of Britain, it appeared, were to cease to be a service for the whole nation and become a concern run on business principles. The surgery Beeching proposed was radical indeed so far as south-west Scotland was concerned. The whole of Galloway was to lose its railway altogether. South-west of a line running through Ayr, Kilmarnock and Dumfries there would be no more train services. Stranraer, in particular, would lose them. No doubt the ferry service would continue; the carriage of private cars and larger transport vehicles would go on; Belfast could still be reached from England by way of Liverpool or Heysham, if one chose not to fly – the quickest and most painless method for those with queasy stomachs, if the most expensive. But the sleeping car rolling grandly through the night from or to Euston would become a thing of the past. Galloway stood aghast at the prospect – Stranraer especially, where many of the railway's employees lived. David Smith, author of *The Little Railways of South West Scotland*, makes comments that are worth quoting:

> There was of course an instant and indignant outcry. I was as indignant as any at this condemnation without adequate reason. Then, in the summer of that year I spent a holiday at Newton Stewart. After that I wasn't so sure. Traffic had dwindled to an appalling extent. Goodness knows I had seen this in many districts; what puzzled me here was the poor patronage of road services also. Yet

Three views taken on 2nd November, 1984. *Top.* No. 47408 leaving Stranraer Harbour station with the 1.30 pm to Glasgow service. Note the motorail terminal and the ferry in the background. *Middle.* The same locomotive seen near Dunragit (note the word 'DUNLOP' on the front of the locomotive). *Bottom.* No. 47016 at Stranraer Harbour having just arrived from Glasgow. *Tom Heavyside*

> the roads were singularly devoid of private cars. I talked with a stationmaster. His views are worth quoting. 'The Galloway people', he said, 'don't need to travel nowadays. The housewife used to go into town twice or thrice a week to shop. Now the travelling shop comes to her door. For entertainment she and her husband went into town to the cinema. Now they have television at their own fireside. And if the young folks want to dance, he's a poor fish who hasn't a motor-cycle or scooter with a pillion seat for his partner. So they just don't use public transport at all'.

In other words, the need so acutely felt a century earlier had now vanished. So far as long-distance travel to Stranraer to connect with the Irish boat service was concerned, however, the threat remained. Then, in February 1965 British Railways put out another Report: *The Development of the Major Trunk Routes.* So far as Stranraer was concerned, this made an important and surprising concession. The line from Ayr to Stranraer would remain, and through trains from Glasgow would operate as before. The former Portpatrick Railway from Dumfries to Stranraer would close permanently as far as Challoch Junction, where it met the line from Glasgow; this would now cease to be a junction. The through services, including the sleeping car service, which had hitherto gone that way, would travel instead through Kilmarnock and Ayr, both of which would now get a through sleeping car service from and to the south. The extra distance to be travelled was considerable, and from Ayr southwards the gradients would be as stiff or stiffer than those on the line about to close. As for the places which the latter had served, express buses would be put on which would only take ten minutes longer between Dumfries and Stranraer, calling at Kirkcudbright (which was also to lose its trains), Gatehouse of Fleet and Newton Stewart.

To this the local inhabitants had to resign themselves, the Stranraer people thinking themselves lucky that they still had a railway. The line from what had been Challoch Junction was to remain as far as the Harbour Station at Stranraer; the latter now became the Burgh's only station, Stranraer Town closing and its associated sheds and outbuildings disappearing, though some lines remained as sidings. Castle Kennedy and Dunragit stations also closed, though trains still ran past their disused buildings. Kirkcudbright, Dalbeattie, Castle Douglas, Gatehouse of Fleet and Newton Stewart went the way of Portpatrick, Wigtown and Whithorn, having only bus services. The change took effect as from 13th June, 1965, the last train being the night sleeper to London the previous evening. Many came to see it arrive and depart, including the Provost of Newton Stewart wearing his chain of office. A century of history had seen the birth and death of a railway, and the mourners went about the streets.

Chapter Twelve

The Route Described

In this chapter the past tense is used throughout, since only six miles of the original line are still in use; the rest has now been dismantled and only earthworks and the occasional remains of bridges, viaducts and other structures are still to be seen. Numbers within brackets are Grid References on the Ordnance Survey 1:50,000 maps of Great Britain; the maps concerned are Nos. 82–84.

The Portpatrick Railway, as built during 1861–1862, extended from its junction with the G&SWR at Castle Douglas for a little over 60 miles to Portpatrick. Except at passing loops and the two chief stations it was single throughout. It was steeply graded, though nowhere rising much above the 500 ft contour; almost a third of the whole distance was a 1 in 100 or steeper. It was therefore not a line for fast running, even after the crossing loops had been altered so that no train running through needed to slow for a reverse curve.

Castle Douglas station (2768 5628) was a sizeable one, with an up island platform from either side of which trains left in the Dumfries direction, and a down platform adjacent to a forecourt where the main station buildings were, from which a short carriage drive led to the western end of the road from the town to Dalbeattie; the main road to Dumfries crossed the railway just west of the station and headed northwards. To the east of the station various sidings fanned out to the south of the line, where there was a goods shed; on the other side, north of the island platform, was an engine shed and a small turntable. A footbridge connected the two platforms. At the western end of these the G&SWR branch to Kirkcudbright bore off to the south-west, the line to Stranraer continuing straight on before veering north-westward. Mile post zero was at this junction.

Bending round the base of the low hill of Dunmuir, the Portpatrick line drew near to the road from Castle Douglas to Ayr (the present A713) and ran parallel with it on its north-eastern side before going under it some 3 miles further on and following the shore of Loch Ken. A little short of 4 miles from Castle Douglas was the station serving the village of Crossmichael (2728 5669), a two-platform structure with the main buildings on the up side, and a crossing loop; beyond the station were two sidings and a long shunting neck. Almost three miles further was the next station at the village of Parton (2691 5701) which had only one platform on the north-eastern side of the line, and no crossing loop; there were three sidings just short of the station.

Immediately beyond Parton the line swung to the west to cross Loch Ken at its narrowest point, where it is scarcely the width of a river. Here was (and still is) the Loch Ken viaduct (*see Appendix 1*); after crossing it the line now curved alongside the Black Water of Dee, a turbulent rocky stream, on the opposite side of which are the ornamental grounds of Hensol House, still standing, whose structure bespeaks the magnificence of its Victorian and Edwardian past, when it was a gentleman's residence. A little further on the main road between Kirkcudbright and New Galloway was reached (the present A762) at the hamlet of Mossdale. Here was New Galloway station (2663 5702) six miles from the small burgh of that name and linked to it in

A busy scene at Castle Douglas station in the late 1890s with a train entering from the Stranraer direction.
Courtesy Ewart Library, Dumfries and Galloway

A fine 1890s view of Castle Douglas station looking towards Dumfries. *Courtesy Ewart Library, Dumfries and Galloway*

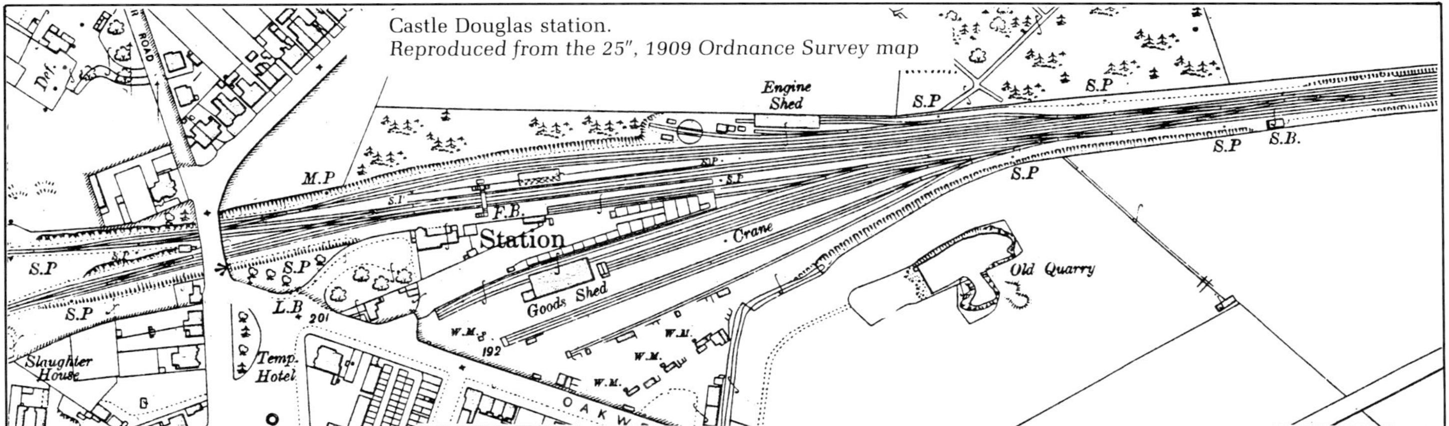

Castle Douglas station.
Reproduced from the 25", 1909 Ordnance Survey map

The Stranraer boat express as seen on a commercial postcard of 1912.
Oakwood Collection

Crossmichael station showing the down platform. Note the oil lamps and the corner of Loch Ken (*right*). Photographed in July 1957. *R.M. Casserley*

A view of Crossmichael station looking towards Castle Douglas with the morning train from Dumfries to Stranraer, headed by a 'Black Five', 4–6–0, No. 5432 entering the station on 4th September, 1961. *H.B. Priestley*

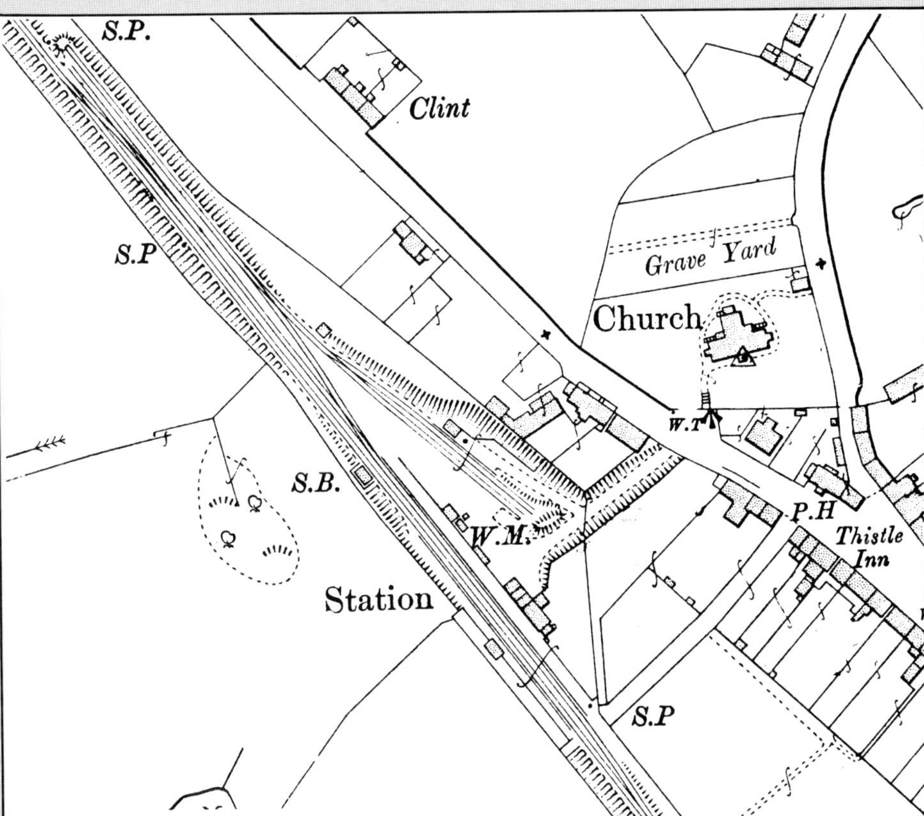

Crossmichael station.
Reproduced from the 25″, 1909 Ordnance Survey map

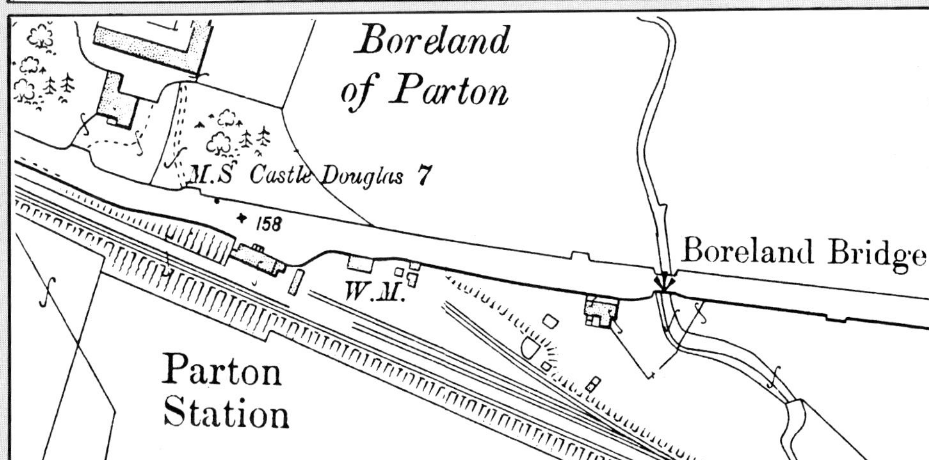

Parton station.
Reproduced from the 25″, 1909 Ordnance Survey map

A turn of the century view of New Galloway station, looking westwards. A double headed passenger train is approaching the 'up' platform headed by what appears to be a Conner 2–4–0, with no spectacle plate and equipped with a stovepipe chimney.

Lens of Sutton

A special train to Stranraer headed by preserved Caledonian Railway 4–2–2, No. 123 seen here at Loch Skerrow on 23rd June, 1962. *R.M. Casserley*

Gatehouse of Fleet station buildings and signal box, situated on the down platform. This little-used station is now showing signs of dilapidation when photographed on 4th September, 1961. This view is looking east at milepost 19.00. *H.B. Priestley*

The line's highest point, Gatehouse of Fleet station captured in the 1930s when in operational condition. *Lens of Sutton*

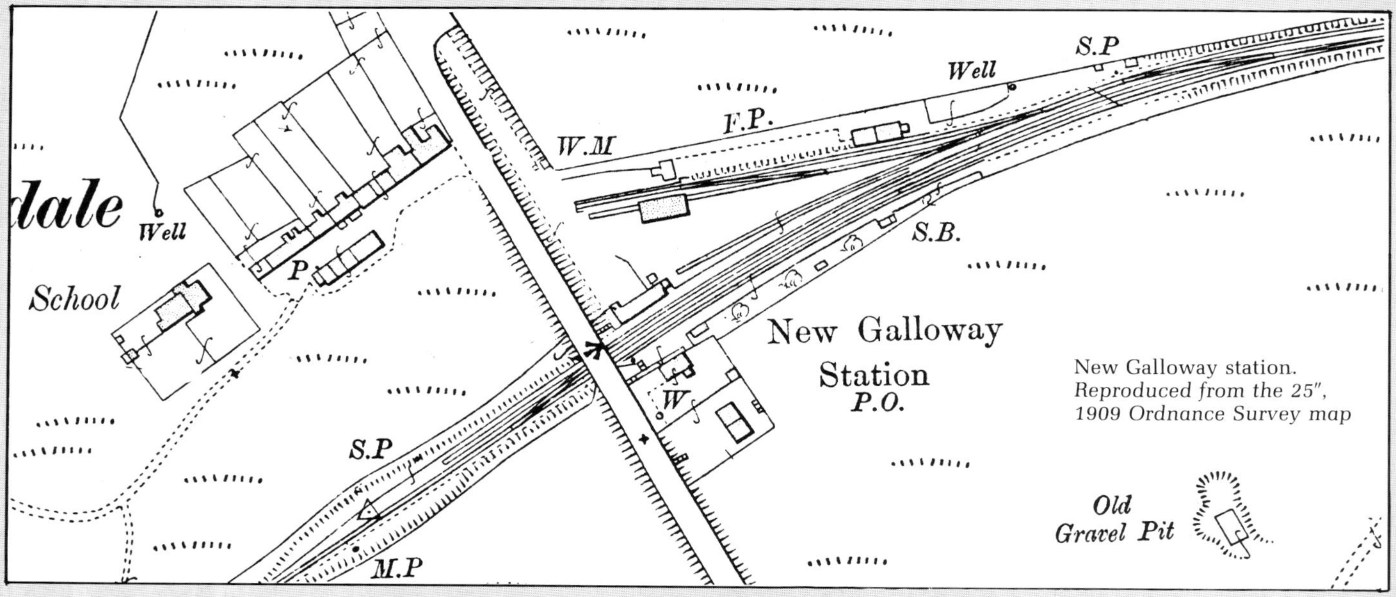

New Galloway station.
Reproduced from the 25",
1909 Ordnance Survey map

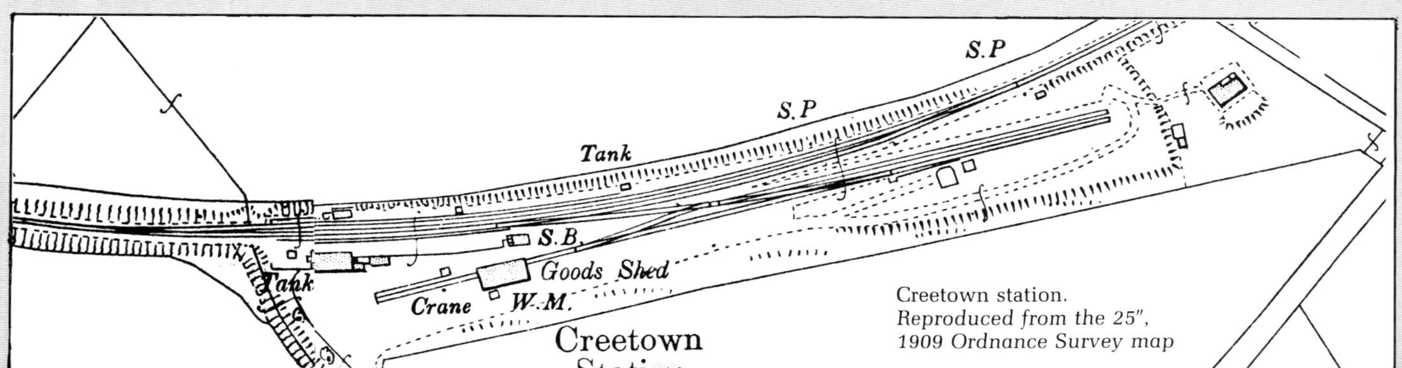

Creetown station.
Reproduced from the 25",
1909 Ordnance Survey map

Creetown station looking north-west. The mid-day service to Dumfries from Stranraer is seen approaching with No. 44993 'Black Five', 4–6–0 on 4th September, 1961.
H.B. Priestley

A further view of Creetown station looking towards Castle Douglas. Note the goods shed by the signalbox and the water tower on the left. *Lens of Sutton*

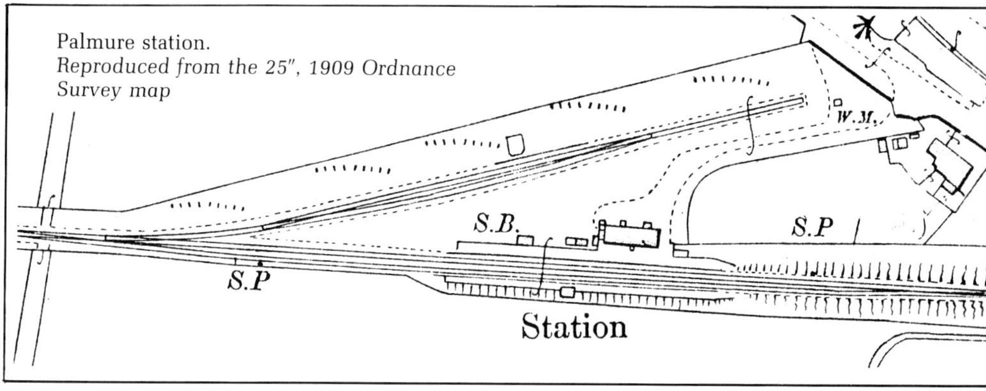

Palmure station.
Reproduced from the 25", 1909 Ordnance Survey map

A post-closure view in September 1961 of Palmure station, with the loop line removed and the whole area very dilapidated. *H.B. Priestley*

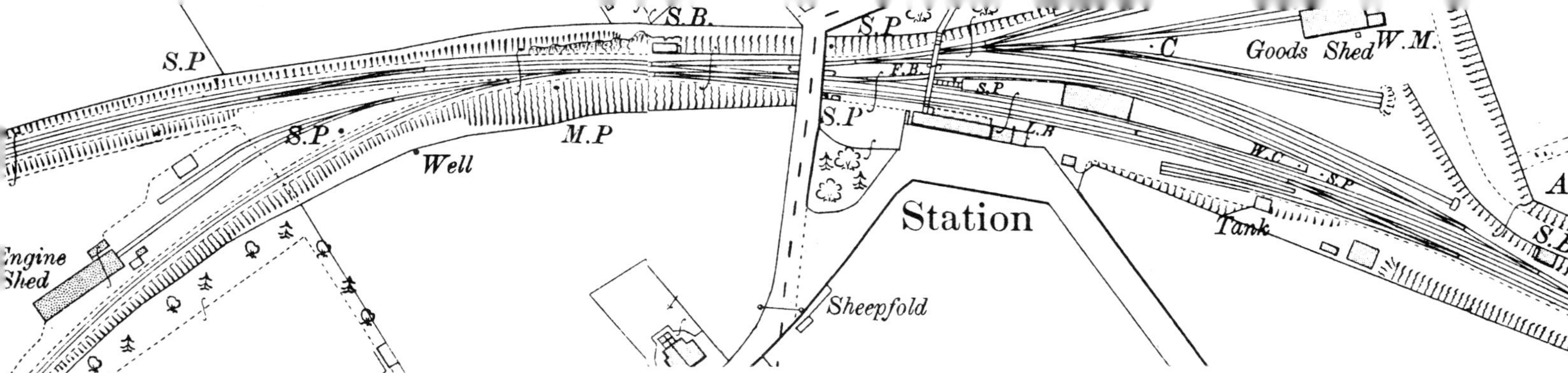
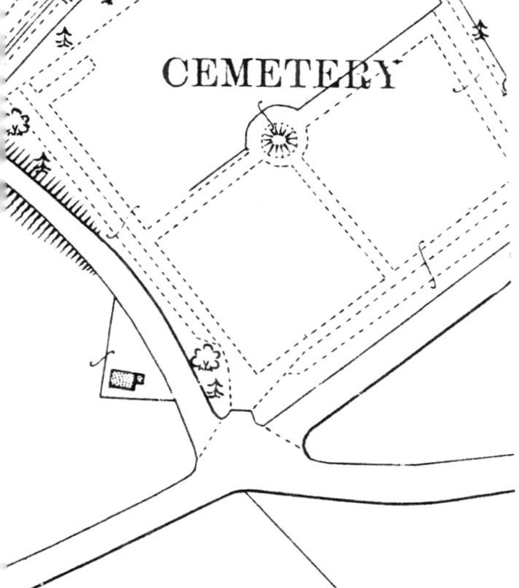

Newton Stewart.
Reproduced from the 25", 1909 Ordnance Survey map

Newton Stewart Junction station, facing Dumfries on 23rd June, 1962 with a train standing at the island platform.

The forecourt of Newton Stewart station on the south side of the station buildings. The signal gantry indicates that the Whithorn's branch diverges just beyond it. Photographed on 23rd June, 1962. *R.M. Casserley*

The up (northern) platform at Newton Stewart photographed from a carriage window on 21st June, 1937. The Whithorn branch trains arrived at (and departed) from the bay behind the station nameboard. *H.C. Casserley*

Kirkcowan station showing the up platform, as photographed on 2nd July, 1957 from a carriage window of a train in the down platform. *R.M. Casserley*

Kirkcowan station looking towards Stranraer. Note the oil lamps with the ladder used for lighting them! *Lens of Sutton*

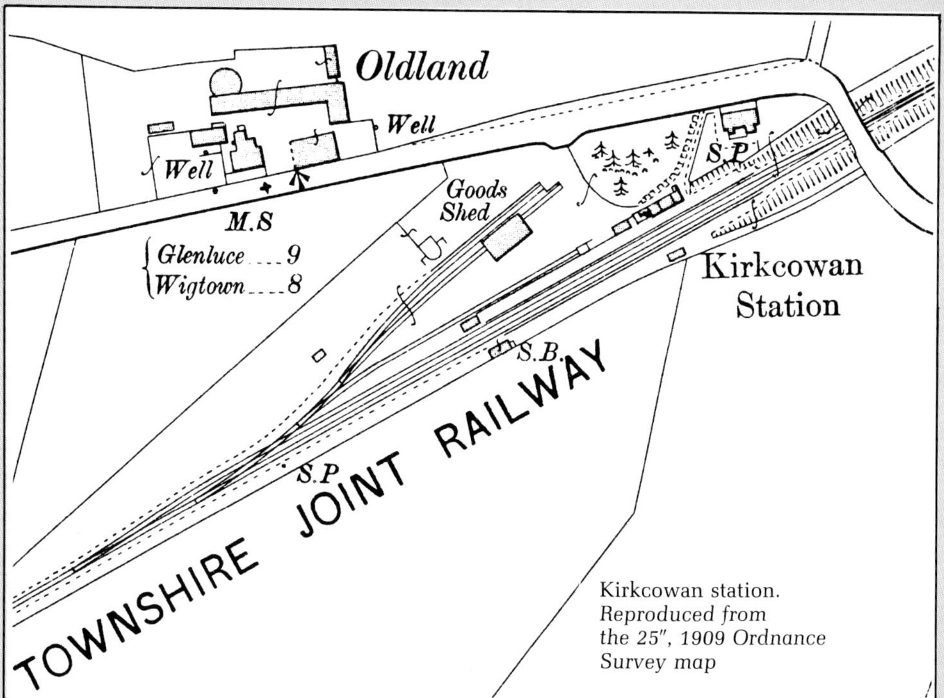

Kirkcowan station.
Reproduced from the 25″, 1909 Ordnance Survey map

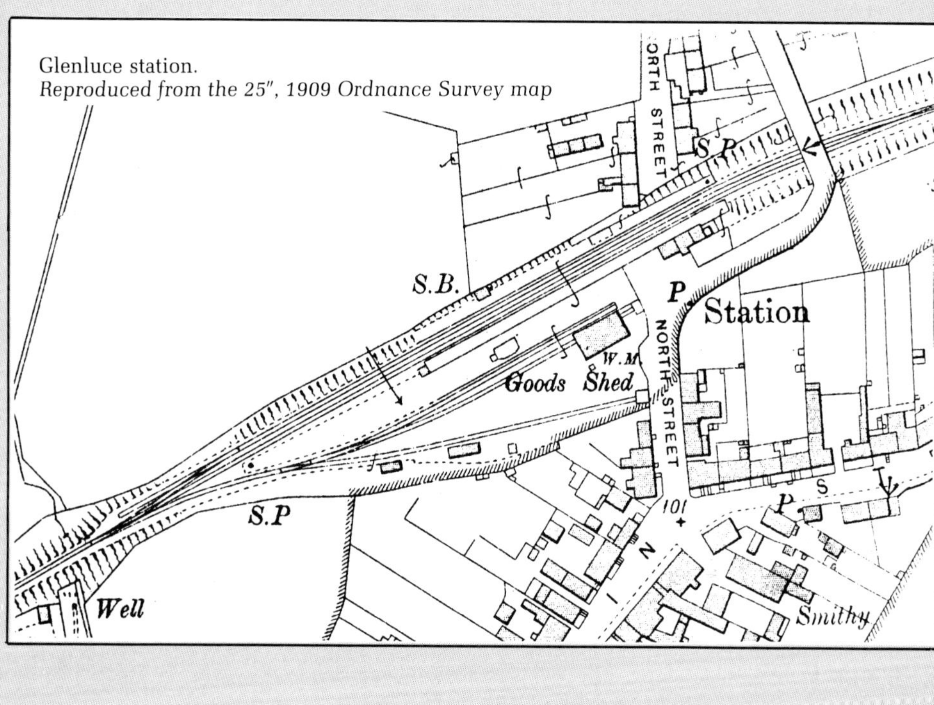

Glenluce station.
Reproduced from the 25″, 1909 Ordnance Survey map

Glenluce station with a view of the up platform and tall signalbox as viewed from the carriage window on 2nd July, 1957. *R.M. Casserley*

No. 40623 Class '2P' 4–4–0 on the up mid-day stopping train seen here crossing another service at Dunragit station on 2nd July, 1957. *R.M. Casserley*

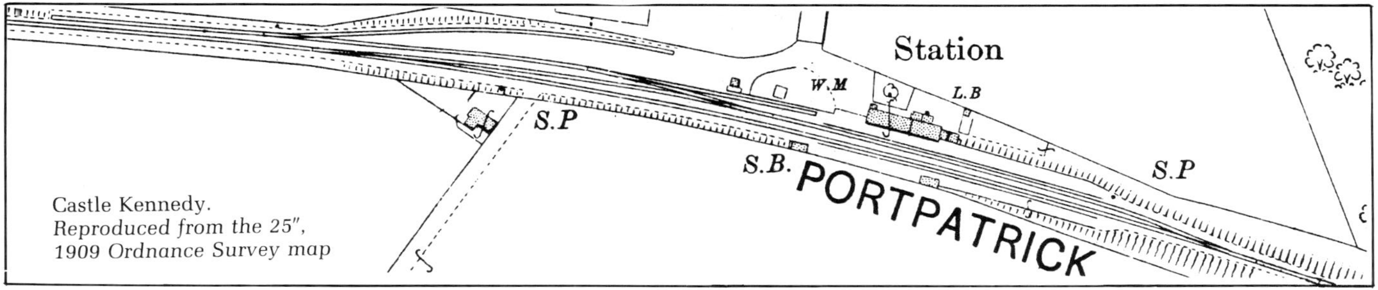

Castle Kennedy.
Reproduced from the 25",
1909 Ordnance Survey map

Drawings of Castle Kennedy station buildings.

Drawn by J.P. Park.
Courtesy Railway Modeller.

Dunragit station in August 1984. This former Portpatrick & Wigtownshire Joint station has been closed for many years but a passing loop has been retained here. Although the booking office (formerly joining the left hand side of the building) has been demolished, the surviving structure retains plenty of 'Scottish' atmosphere. *S.C. Jenkins*

Dunragit station and goods yard as seen from an approaching train (7.52 am Stranraer to Glasgow) on 9th May, 1959. The service was being hauled by BR Standard Class '5', No. 73101 (shedded at 67A). *H.B. Priestley*

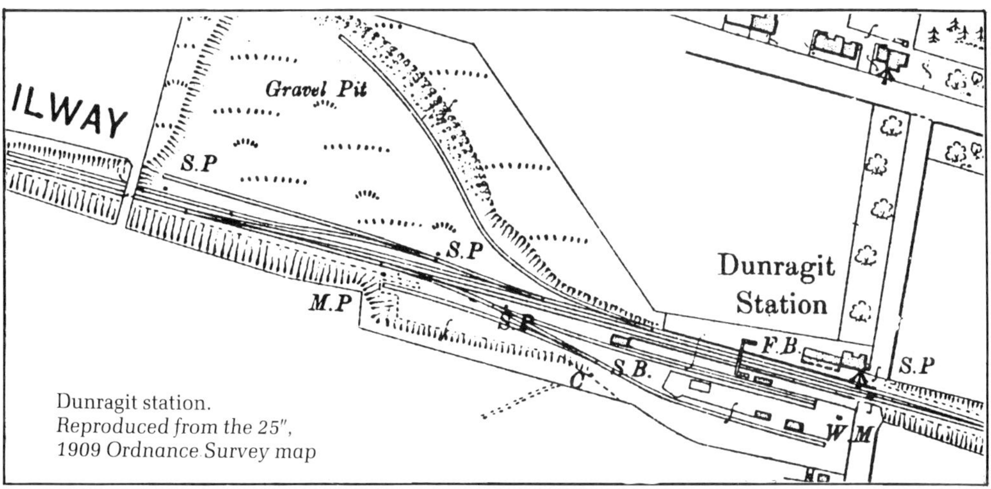

Dunragit station.
Reproduced from the 25", 1909 Ordnance Survey map

Stranraer Harbour station with a 'Black Five' 4–6–0 No. 45384 seen here with the up boat train to Glasgow. This photograph was taken on 12th June, 1964 before the reconstruction.

H.C. Casserley

No. 57375, 0–6–0 'Jumbo' class (formerly Caledonian Railway No. 201) seen here on 23rd June, 1962 with a RCTS/SLS special Scottish Railtour at Stranraer Harbour station.

H.C. Casserley

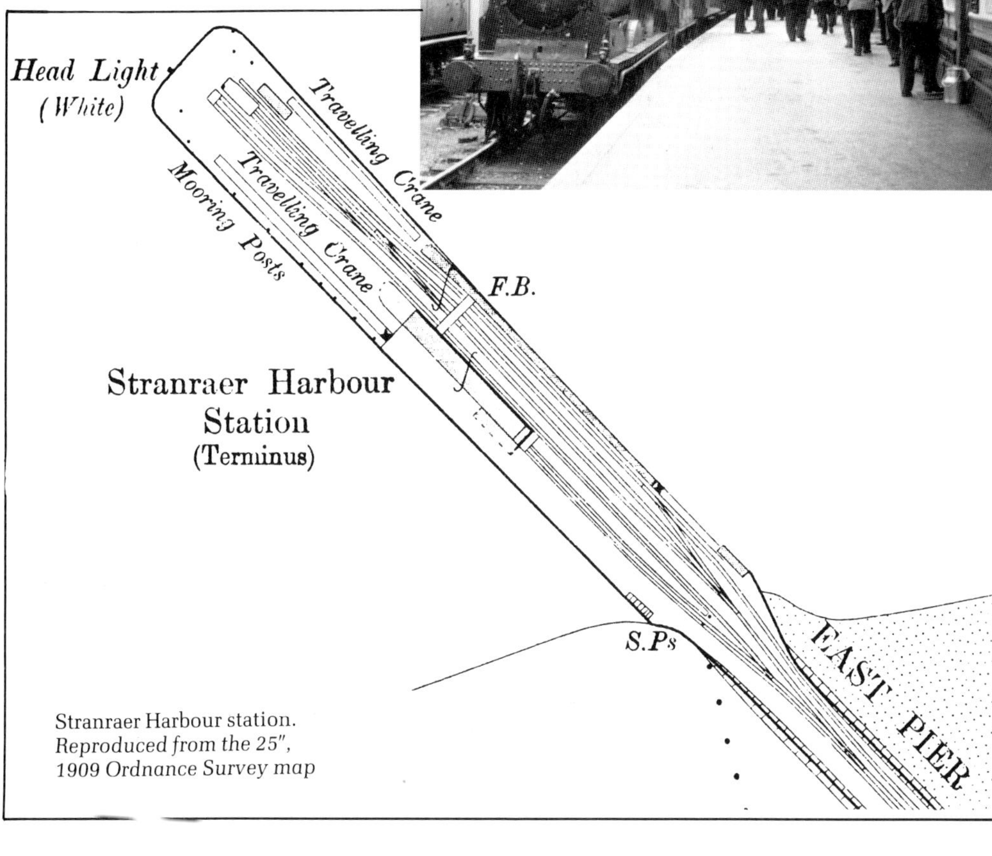

Stranraer Harbour station.
Reproduced from the 25", 1909 Ordnance Survey map

Midland Railway, three cylinder compound No. 1179 seen here in 1937 on the turntable at Stranraer. These locomotives were generally used on the Glasgow to Stranraer service (via Girvan) and rarely on the line to Castle Douglas. *H.C. Casserley*

LMS 0–6–0 'Jumbo' class No. 17440 of Drummond design, seen here rebuilt with new boiler and stovepipe chimney, at Stranraer on 23rd April, 1948. Note the provision of a 'fender cab' for tender-first running. *H.C. Casserley*

LMS Class '2P' No. 600 seen here at Stranraer Town on 23rd April, 1948 (now in BR ownership). *H.C. Casserley*

The 11.53 Sprinter from Glasgow having just arrived under the covered roof of Stranraer Harbour station in July 1989. The main platform (the only one now regularly used) can be seen in the lower photograph, the other platforms (very occasionally used) having no weather protection. Note the device beneath the buffers intended to deal with the snow and small objects on the line. *Author's Collection*

The very busy locomotive sheds and workshops at Stranraer seen here at the turn of the century. *Courtesy Railway Magazine*

The water towers and locomotive sheds at Stranraer seen here on 23rd April, 1948. The engines are (*left*) 2–6–0, 'Crab', No. 2905 and 4–6–0 'Black Five' No. 4484. *H.C. Casserley*

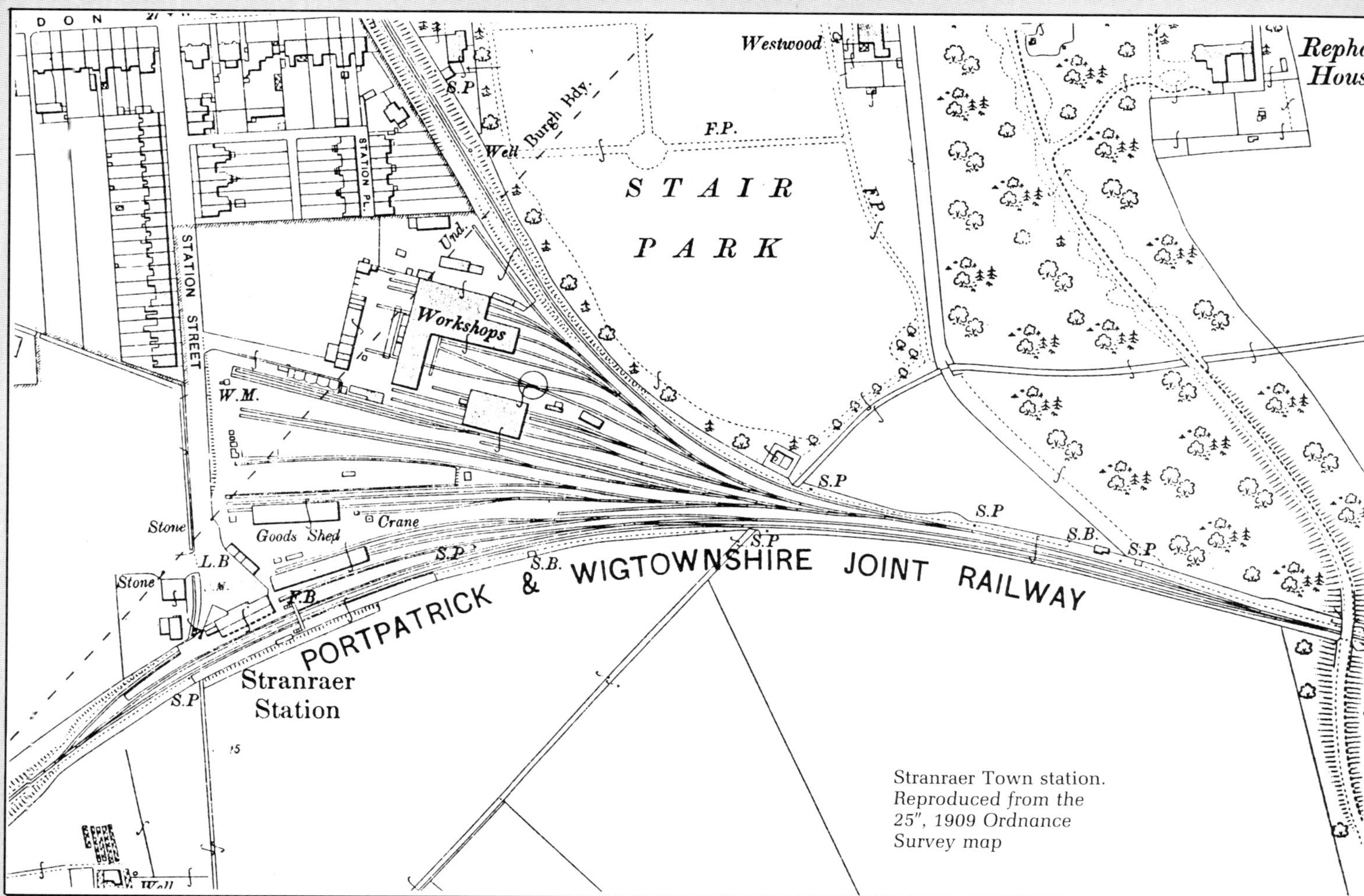

Stranraer Town station.
Reproduced from the 25", 1909 Ordnance Survey map

LMS 'Black Five', No. 4884 in a very grubby condition about to leave Stranraer Town station on 23rd April, 1948. The schoolboy spotter (wearing the kilt) watches the departing service for Dumfries with awe. *H.C. Casserley*

'Jubilee' class, 4–6–0, No. 45704 (shed 12A) seen here on 9th May, 1959 at Stranraer Town station after bringing the night boat train from Euston to Stranraer Harbour. Also in view is BR Standard Class '5' No. 73101 at the head of the 7.25 am morning service to Glasgow (via Girvan). *H.B. Priestley*

Stranraer Town station showing signal (cut-down) and signal box in BR days.
Lens of Sutton

Again at the other end of the platform, signals utilising a cut-down post as support for the bracket.
Pendon Museum

LMS three cylinder 4–4–0 compound No. 916 has just arrived at Stranraer Harbour with a train from Carlisle on 21st June, 1937.
H.C. Casserley

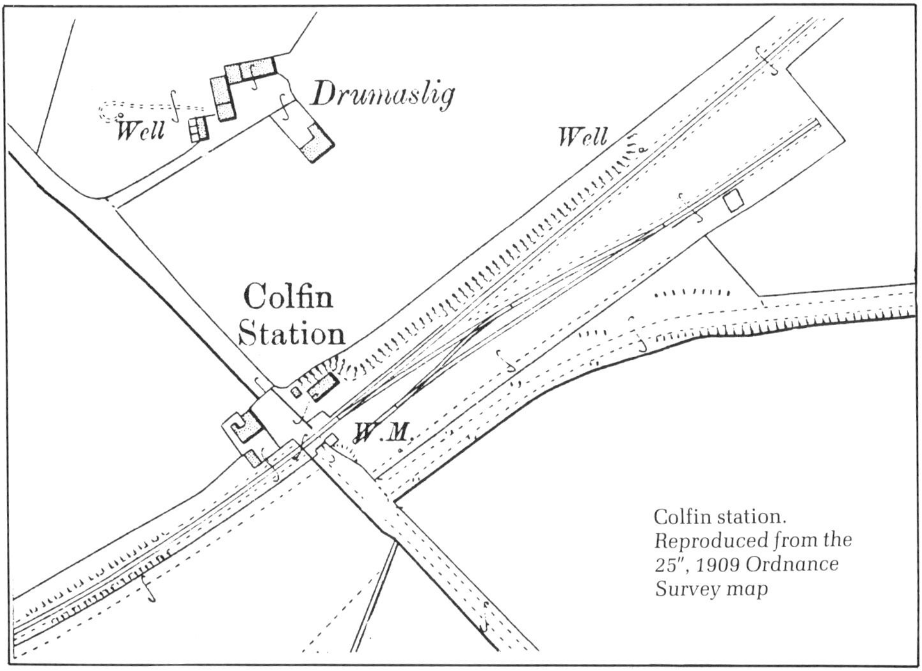

Colfin station. *Reproduced from the 25", 1909 Ordnance Survey map*

Portpatrick station, showing the branch passenger train headed by ex-Caledonian Railway 0–6–0 locomotive No. 17440. The harbour branch formerly diverged to the left through the present site of the platform from a point between the two station buildings. *D.S.M. Barrie*

LMS Class '2P' 4–4–0 No. 645 has just arrived at Portpatrick on 21st June, 1937 with an afternoon train from Stranraer. When returning it will work tender first as Portpatrick was not equipped with turntable. *H.C. Casserley*

The stone water tower and early engine shed (now disused) at Portpatrick station photographed on 21st June, 1937. *H.C. Casserley*

the days of horse-drawn traffic by a coach service which took ¾ hr. The station had two platforms and a passing loop; the main building was on the up platform, behind which were three sidings and a small goods shed.

So far the line had run through pleasant green lowland country with scattered farms and villages, but from New Galloway station onwards it traversed a wild and uncultivated area for a considerable distance, a table-land dissected by mountain streams and spreading out south eastwards from the highest hill in this area, the Cairnsmore of Fleet (2,331 ft). The aspect of this region was mostly barren and treeless when the line was built, though it has since been the scene of much afforestation. The line rose steeply, mostly at 1 in 80, for four miles, crossing the Black Water of Dee a little over a mile beyond the station by a large four-arch stone bridge and then curving round the base of Airie Hill to run southward along the eastern shore of Loch Skerrow, a featureless sheet of water formerly, though backed by pine trees now. Here was a passing place with a siding, a signal box and a water tower into which water was pumped by a stationary steam engine, to replenish the tanks of locomotives which always found the climb up from Loch Ken a thirsty business. A rough track led off to the south which after five miles reached a metalled road that led down to Gatehouse of Fleet; otherwise there was no means of access to this remote spot except along the line itself.

From Loch Skerrow a mile and a quarter of level track preceded a ½ mile dip at 1 in 76 to cross the Little Water of Fleet by a nine-arch viaduct (now destroyed) and a further mile and a quarter up at the same gradient was followed by a one mile fall to the Big Water of Fleet viaduct, a massive structure of 20 arches, visible for miles around, and at the moment of writing (early 1990) still standing, though in a precarious state. Its piers were reinforced early in the present century with sheaths of brickwork. A little further on, to the right, are the Clints of Dromore, a fine-looking array of crags with a gash through their centre, which rise to almost 1,000 ft above sea level. Approaching their eastern end the line turned southwards, still rising steeply, and reached its summit at the small station originally named Dromore, but later re-named Gatehouse of Fleet (2546 5625) which town was actually six miles away to the south-east at the lower end of the Water of Fleet. The station had two platforms and a passing loop; close by was the station master's house. A narrow road led down the west side of the valley to Gatehouse itself. During the early 1900s a coach service connecting with such trains as stopped there took an hour and twenty minutes to descend to Gatehouse, and (naturally) a good deal longer to get back up there again.

The track now fell steeply on an almost continuous slope of 1 in 80 practically to sea level, first following the line of the Moneypool Burn and then curving round to the north above Creetown village. Creetown station (2472 5600), from which a magnificent view could be had of Wigtown Bay and the Machars to the south and south-west, had two platforms and a crossing loop; the station buildings were on the south platform; on this side, too, were sidings and a small goods shed. The descent continued to cross the Palnure Burn by a three-span girder viaduct and, turning to the west, reached Palnure station (2452 5634) just across the Gatehouse–Newton Stewart road (the present A75). Here were the customary two platforms and a crossing loop, with the station buildings on the north platform at the end of an approach drive from the road; on the same side were sidings in the form of a long loop with a shunting neck beyond. A mile further on the river Cree was crossed just before its sinuous lower course widens into an estuary, by a three-span girder bridge supported on timber piers. Now followed a fairly steep rise, at first westward along an embankment built on a wide green haugh looped round by the Cree, the Carse of Barr; the road from Whithorn to Newton Stewart (the present A714) was then crossed and the line turned north-west again along a hillside, from which there was a splendid prospect towards the Cairnsmore of Fleet and other high Galloway hills. Three miles beyond the Cree Bridge was Newton Stewart station (2405 5650).

This was the largest station between Castle Douglas and Stranraer, and just under 30 miles from the former. Newton Stewart was Galloway's second largest town; the station had a large island platform on the up side and a single down platform whose main buildings opened on to a large forecourt. The town itself lay to the north-east. The down platform had a bay with a shunting neck facing towards Castle Douglas; north of the island platform there was an extra through line enabling the station to be avoided altogether, and from its western end six sidings fanned out eastwards, one of which went through a goods shed. The island platform was partly covered with a glass canopy roof extending its whole width. From the western end of the station the branch to Wigtown and Whithorn diverged sharply to the south, and in the angle between it and the main line an engine shed provided cover for the locomotives used on that branch. There was no turntable for them, so locomotives proceeding to Wigtown or Whithorn had to travel bunker or tender first in one direction.

The line to Stranraer and Portpatrick continued, generally running parallel to the main road to these places in a west-south-westerly direction, climbing fairly steeply for 2½ miles and then falling again to the bridge across the river Bladnoch; it then veered away from the road to the south-west to serve the village of Kirkcowan. The station (2322 5610) was a ¼ mile from the village to the west; it had the usual two platforms and passing loop, with the main buildings on the up side, where there were also three sidings and a goods shed. Continuing more or less parallel to the main Stranraer road, but a mile to the south, the line crossed the Tarf Burn 2½ miles further on by a substantial bridge, converged once more upon the road, crossed it, ascended to a summit about 40½ miles from Castle Douglas and then descended fairly steeply to Glenluce, four miles further on. The station here (2199 5577), had two platforms with a passing loop, and the main buildings on the down side, where there were also three sidings and a goods shed. Building the station involved blocking one of the roads in the village and making a divergence for it across a bridge just to the east.

A little further on was one of the most striking viaducts on the line, which is still standing; here the railway was carried high above the Water of Luce just a mile from its mouth. Unusually in this part of Scotland, the viaduct was built of red stone; it has eight arches, under which the road to Stranraer west of Glenluce village loops itself, going under one arch, crossing the

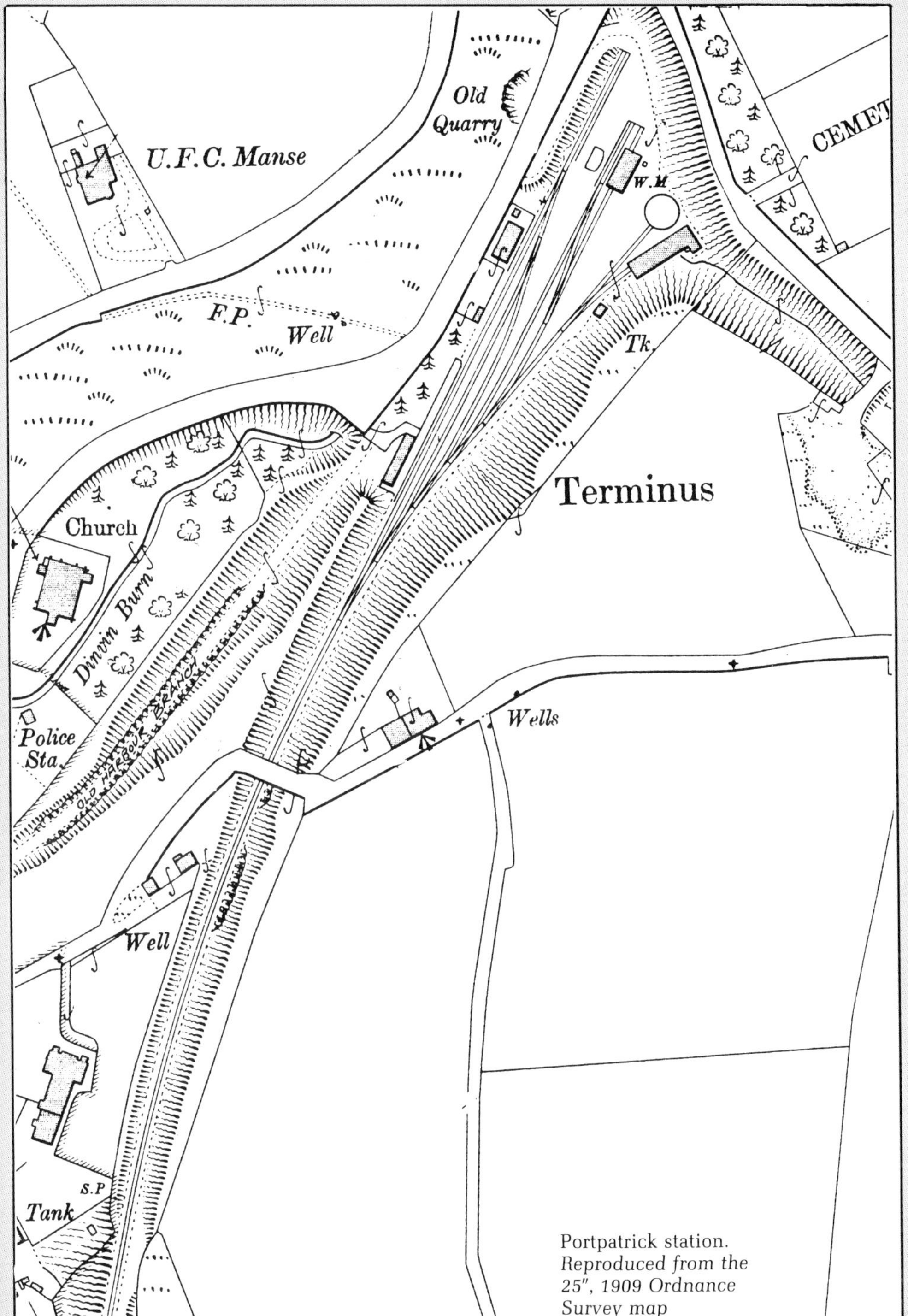

Portpatrick station. Reproduced from the 25", 1909 Ordnance Survey map

stream by its own bridge and doubling back under another arch. The line, meanwhile, was still falling, and reached sea level again at Dunragit after the railway from Glasgow and Ayr had curved in from the right to join it at Challoch Junction.

Dunragit station (2153 5573), like most of the other minor stations on the line, consisted of two platforms with a passing loop, the main building being on the up side. It had, however, sidings at both sides, two with a shunting neck to the south of the down platform, and three more to the north of the up platform and a little beyond it, one of them serving a large gravel pit. There was no village at Dunragit; the station probably owes its existence to the fact that one of the line's chief promoters, Sir James Dalrymple Hay, had his mansion less than ½ mile to the north. Another three miles up a slightly rising gradient brought the line to Castle Kennedy station (2108 5597), which also had two platforms and a passing loop, two sidings and a shunting neck. Here, too, one may suspect that there would have been no station had not the Earl of Stair lived close by at Castle Kennedy itself.

A mile and a half further on a steep descent at 1 in 86 began to Stranraer Harbour Junction; the short branch to the East Pier now bore off to the right and in another ¼ mile Stranraer Town station (2067 5604) was reached. This was the largest station on the line, and the headquarters of the Portpatrick and Wigtownshire Joint Committee, where the company's administration was based and where workshops undertook minor repairs to locomotives and rolling stock. Many sidings fanned out into the angle between the main line and the Harbour branch.

Stranraer was a busy place. The Town station had two large platforms, the up one having a bay at either end and being considerably longer than the down platform; the main buildings abutted on to it and a fine glass canopy covered it at this point. It was central for the town itself, unlike that at Newton Stewart. It was also the terminus for all trains from the east except the through boat trains, though through running to Portpatrick had originally been intended in the days when it was still supposed that this would be the point of departure for Ireland, so the rails ran right through the station and continued with what was in effect the Portpatrick branch.

The latter was very steeply graded. Directly it left Stranraer it began to climb at 1 in 72 for four miles, running at first parallel to the main road from Stranraer to Drummore (the present A716), then crossing it and curving round to the south-west, passing through a row of cuttings on to an embankment which led to the last of the line's many viaducts, the one over the Piltanton Burn. Curving south and then south-west again, and crossing the Stranraer to Portpatrick road, it reached its summit at Colfin station (2039 5548), a halt with a single platform and no passing loop but having two sidings with a shunting neck between them. The actual hamlet of Colfin was some distance away to the east, but a large creamery was built beside the station, which probably derived most of its custom therefrom. The final descent now began, half a mile slightly downhill steepening to 2½ miles at 1 in 65 to 1 in 57, curving first to the west, then again to the south-west along the line of a burn and apparently making straight for the cliff edge near Dunskey Castle. Here the line made a complete semicircle to face north-east and finish high above the Burgh of Portpatrick at a single platform with an over-run beyond (2004 5546).

The station site had had to be excavated out of the cliff face, where room was found for six sidings together with goods and engine sheds. There had originally been a branch to the harbour, to reach which trains had to go forward into a very short shunting neck and then back very steeply down to the harbour basin. However, the failure of packet boats to call at Portpatrick meant that this spur was redundant, and after a while its rails were lifted.

THE WHITHORN BRANCH

The branch from Newton Stewart to Wigtown traversed pastoral country and reached no high altitudes, but none the less was quite heavily graded, almost a third of its length being at 1 in 100 or steeper. After a more or less level mile, in which it curved away to the south from the main line at Newton Stewart, it fell nearly to sea level at Causewayend siding and then remained almost level as far as Wigtown, skirting round the low hill on which this burgh stands; Wigtown station (2435 5549) was ¼ mile south of the town on the road to the harbour at the mouth of the river Bladnoch. A long approach drive extended along the west side of the line to the single station platform (on this branch all the stations except Millisle Junction had one platform only) where there was a passing loop. On the other side of the loop there were three sidings, one of which passed through a goods shed, and a loading bay. Here, when the line had been run as an independent concern, had been the workshops where repairs to engines and rolling stock were carried out.

Beyond Wigtown the line crossed the Bladnoch near its mouth by a two-span girder bridge and then turned southward again, over two embankments with a cutting between; here one had a good view across Wigtown Bay with (during the 1940s) the intermediate prospect of a large airfield with two runways, used by the Royal Air Force. At the point of crossing of the Wigtown–Whithorn road was Kirkinner station (2421 5517) where the passing loop was some distance beyond the platform; diverging from it were two sidings and a loading bay. The line now rose steeply to reach a summit at Whauphill station (2404 5499) where the crossing loop had the single platform at its west end and, behind the latter, two sidings and a goods shed. At Whauphill coach connections were made with Port William, seven miles away on the shore of Luce Bay; horse-drawn coaches connecting with trains in pre-World War I days took from one to one and a half hours to make the journey.

From Whauphill the line fell fairly steeply through broken country through a number of cuttings towards the south-east. The next station, Sorbie (2435 5476) was ½ mile north of the village of that name and close to a road junction; the platform and station buildings were on the south-west side of the line and a large creamery lay to the north. The passing loop here was very short, and from it a single siding branched off on the south-west side. About two miles further on, almost at the foot of a declivity, was

Kirkinner station looking south, with the loading dock and bay beyond the main platform; photographed in the early thirties. *Lens of Sutton*

Whauphill station looking north towards Newton Stewart, a single platform with oil lamps; note the abundance of milk churns from the nearby creamery. Again photographed in the thirties. *Lens of Sutton*

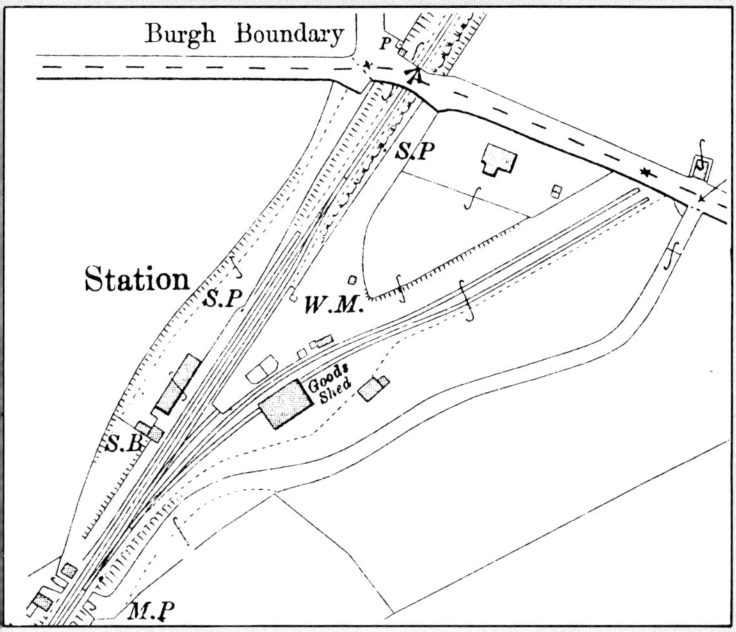

Wigtown station. *Reproduced from the 25", 1909 Ordnance Survey map*

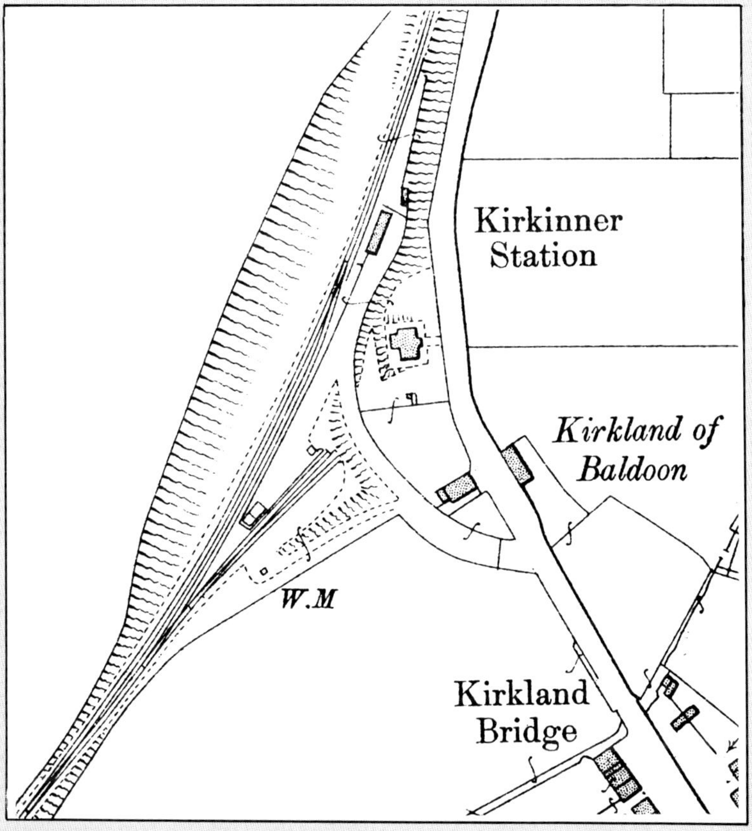

Kirkinner station. *Reproduced from the 25", 1909 Ordnance Survey map*

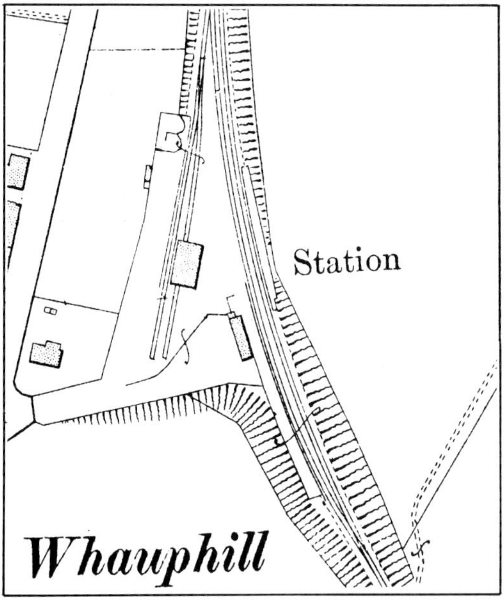

Whauphill station. *Reproduced from the 25″, 1909 Ordnance Survey map*

The distinctive station buildings at Sorbie, on the Whithorn branch of the P&WJR. The right hand building was the station proper, while the adjoining single storey wing at the rear provided accommodation for the station master and his family.
S.C. Jenkins

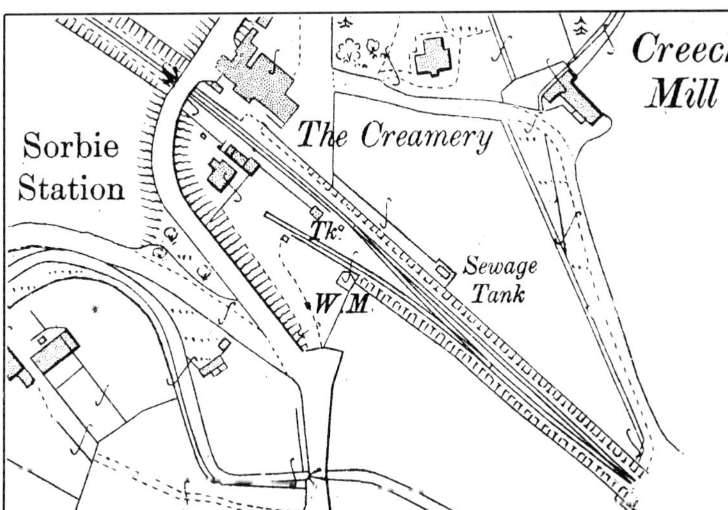

Sorbie station. *Reproduced from the 25″, 1909 Ordnance Survey map*

Steam railmotor standing at the branch station of Garliestown in an early view of this branch (*see map page 102*). *Heyday Publishing Co. Ltd*

Millisle (for Garlieston) Junction, the only two-platform station on the Wigtownshire Railway. A view taken in 1933. *Lens of Sutton*

Garliestown station.

Reproduced from the 25", 1909 Ordnance Survey map.

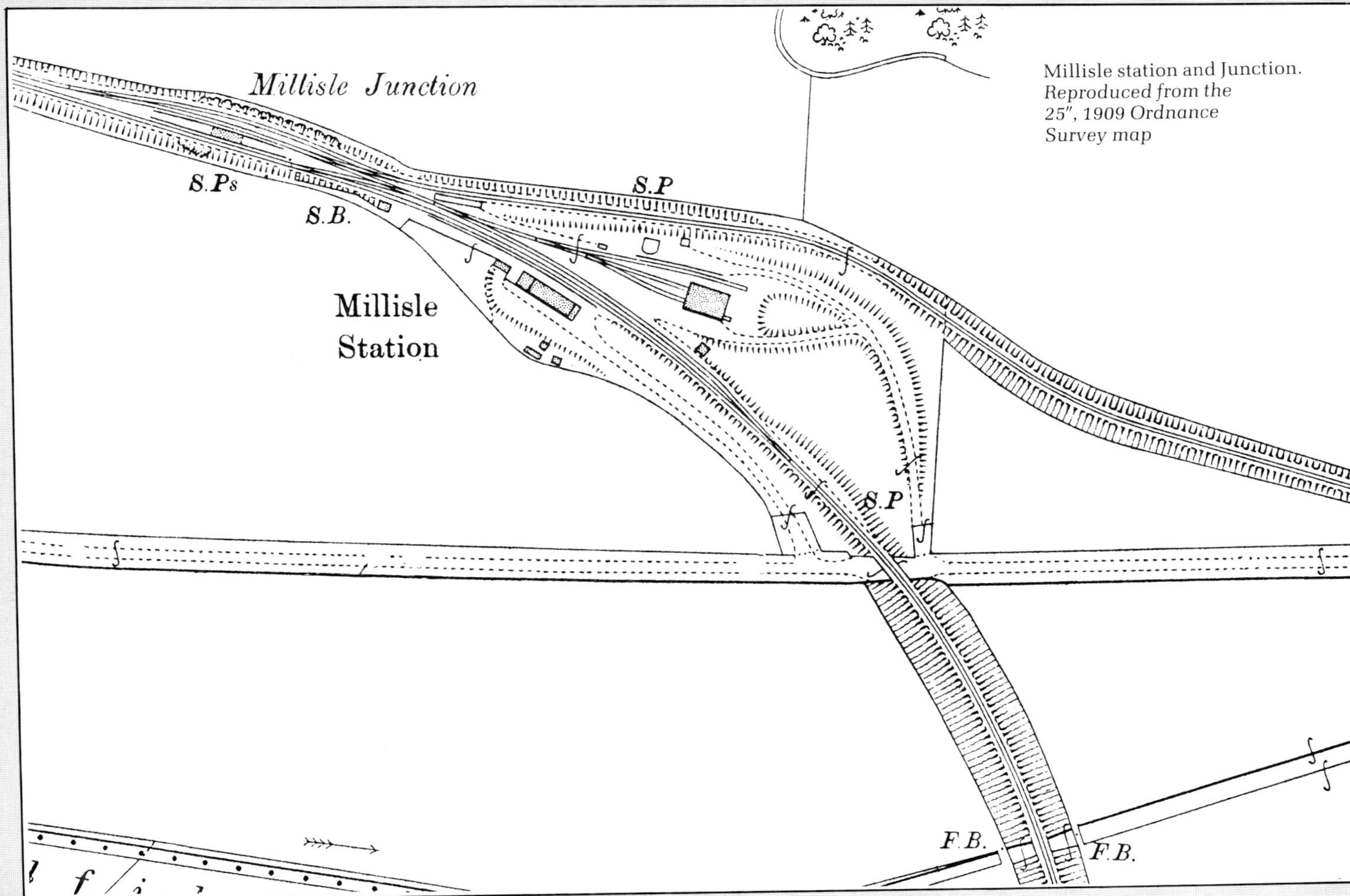

Millisle station and Junction.
Reproduced from the
25", 1909 Ordnance
Survey map

Preserved Caledonian Railway 4–2–2 No. 123 and the Great North of Scotland Railway 4–4–0 No. 49 *Gordon Highlander* seen here at the head of a special train about to leave Whithorn on 23rd June, 1962. The first coach is one of the two preserved coaches repainted in CR livery. *H.C. Casserley*

Whithorn station seen here in its original state, with platform awning. This being the southernmost station in Scotland. *D. Rokeby*

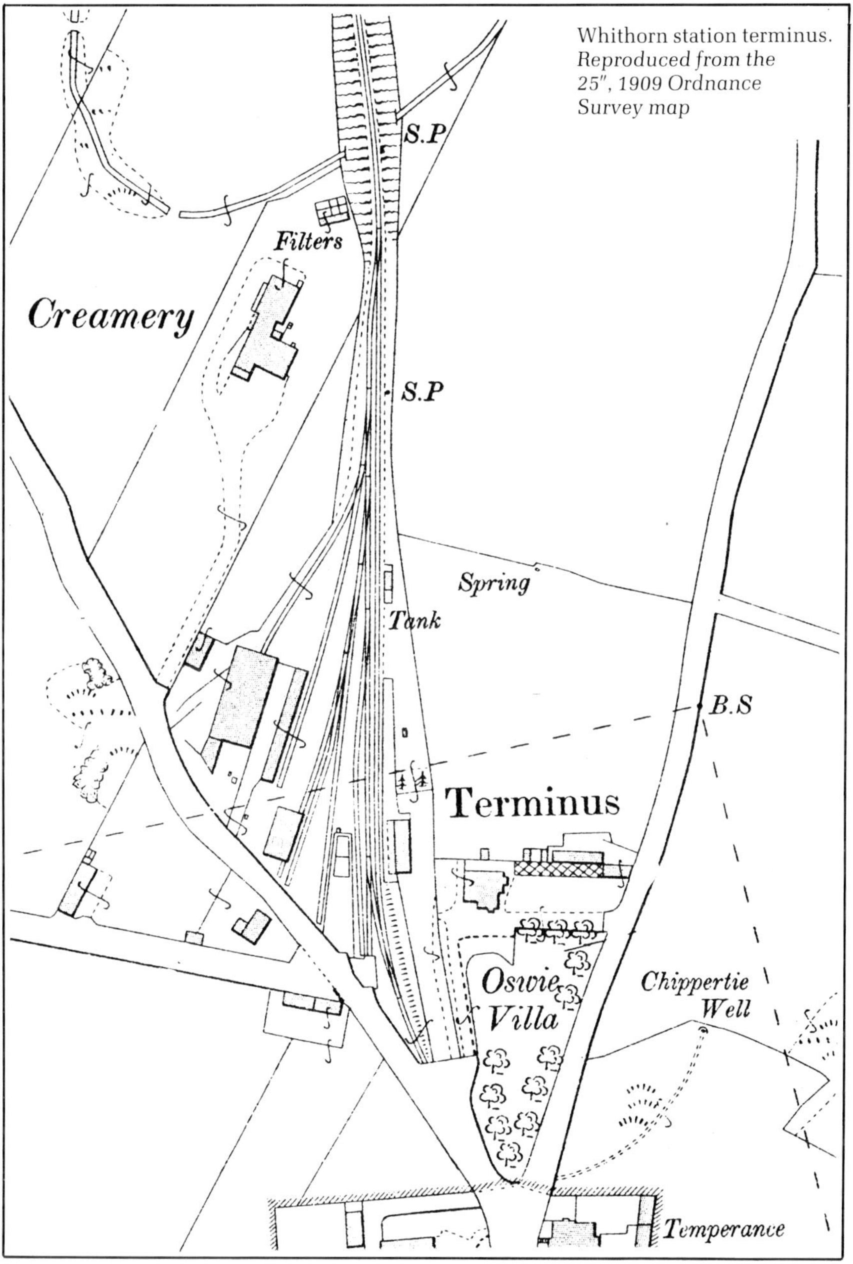

Whithorn station terminus. *Reproduced from the 25", 1909 Ordnance Survey map*

Millisle Junction (2463 5464) where the short branch to Garlieston diverged. This latter continued south-eastwards and then turned eastwards, passing south of Garlieston village, where there was a platform and loop, and then continuing as far as the pier (2481 5462) at which quite large vessels were able to berth at high tide. From the harbour, in clear weather, one may see and perhaps even identify some of the higher hills of the English Lake District, 50 miles distant across the Solway Firth. Millisle station itself lay just beyond the junction; its platform was midway along the south side of the passing loop; in the angle between the branch and the loop were three sidings and a shed for the locomotive which worked the branch. There was no village at Millisle – it was named after a nearby farm – and little traffic could have originated there.

The line now curved sharply, first to the south, then to the south-south-west, and began the final climb, largely at 1 in 60 and 1 in 58, to the summit (and highest point on the branch) a mile short of Whithorn. The station there (2446 5408) lay in the angle between the main road from Whithorn to Wigtown and the minor road to Garlieston. The platform and main building were to the east of the long crossing loop, from whose western side no fewer than six sidings fanned out, on two of which were an engine shed and a warehouse; close by was the inevitable creamery. Thus one arrived at one of Scotland's smallest burghs, which has claims to be the oldest, since it was here that St Ninian founded his monastery, the 'White House' (*Candida Casa*) in the 5th century AD. Whithorn station has its own claim to particularity; it was the southernmost railway station in Scotland. From it, connecting with the trains in the line's palmier years, horse-drawn coaches plied to Isle of Whithorn, the fishing village at the southern tip of the Machars from which the Isle of Man can be clearly seen and from whose harbour occasional boats went across the intervening 30 miles to Ramsey in the summer season. The coaches took ¾ hr for the four mile journey.

Whithorn station after the removal of the station awning. Again this was photographed on 23rd June, 1962 when the special visited the now somewhat dilapidated branch (traffic for passengers ceasing in 1950). *H.C. Casserley*

Chapter Thirteen

Locomotives and Rolling Stock

As already mentioned, the Portpatrick Railway when it first opened ran its own train services between Castle Douglas and Portpatrick with its own locomotives, bought from an engineering firm or (in one case) from another railway. The first three, ordered from Sharp, Stewart and Co. of Manchester, were mixed-traffic 0–4–2s built to the design of James McConnell, locomotive superintendent of the Southern section of the LNWR. Pictures and details of these are not available, but they were probably very similar to the standard type which this firm built from 1848 onwards for nearly 40 years and sent to various companies in Great Britain and Ireland. Two were delivered at the end of 1860 and the third came just in time for the opening of the line. In addition a LNWR goods locomotive of Ramsbottom's design was hired from that company and later, in the summer of 1861, bought outright. The locomotives were numbered 1–3 and (the ex-LNW engine, an 0–6–0) 5. The following year another 0–4–2 was supplied by Sharp Stewart and numbered 4.

Two 2–2–2s, also of McConnell's design, and also built by Sharp Stewart, did not arrive until the end of the summer. They probably resembled the 'Small Bloomers' which at that time were operating so successfully on the LNWR. Thus, by the end of 1861 the Portpatrick Railway had seven engines, two passenger, four mixed traffic and one goods. Meanwhile its newly-appointed locomotive superintendent, Alexander Walker, was considering an eighth. He seems to have been impressed by the possibility of obtaining one with a leading bogie, and made a report on that subject to the Board, which agreed in principle that such a locomotive would be worth obtaining. Sharp Stewart were then asked to give an estimate on the basis of a diagram submitted to them. What happened next is not certain, but the firm seems not to have thought it a good idea; very few such locomotives were then in existence; a front bogie was a very new-fangled thing. At any rate, the engine actually delivered to the Portpatrick Railway during 1863 was a 2–4–0. It was the last one to be built for it during its short period of independent operation and was numbered 8.

With this miscellany of engines the company had to run all its trains, but once the CR took over the running of the line in 1864 others were added, and gradually the first eight were replaced as they wore out. The operating company brought in one by one its very successful 2–4–0s built to the design of Benjamin Conner, the CR's locomotive superintendent from 1857 until he died in office in 1876. He constructed some 200 with this wheel arrangement, some with 7 ft driving wheels, the rest with 6 ft 2 in. wheels, and some were sent to the Portpatrick line. A coloured picture of one, hauling a train of 4-wheeled coaches and painted in bright Caledonian blue, appears as the frontispiece of David Smith's book, *The Little Railways of South West Scotland*. They were outside-cylindered engines and had 4-wheeled tenders holding only 1,500 gallons of water, so that frequent stops for re-filling were necessary. They had stovepipe chimneys, domes over their fireboxes when first built (which were subsequently moved to the more conventional position over the boiler), slotted wheel splashers and pierced

Glasgow and South Western 'Smellie', 4–4–0 No. 14116 with 6 ft 1 in. driving wheels as used on the Portpatrick line after 1885, seen here in LMS livery. *H.C. Casserley*

One of Drummond's superheated 4–4–0s, with 6 ft 0 in. wheels which worked the services between Dumfries and Stranraer with great success, during the First World War but provided too heavy for the permanent way and had to be withdrawn from the line. *Courtesy National Railway Museum*

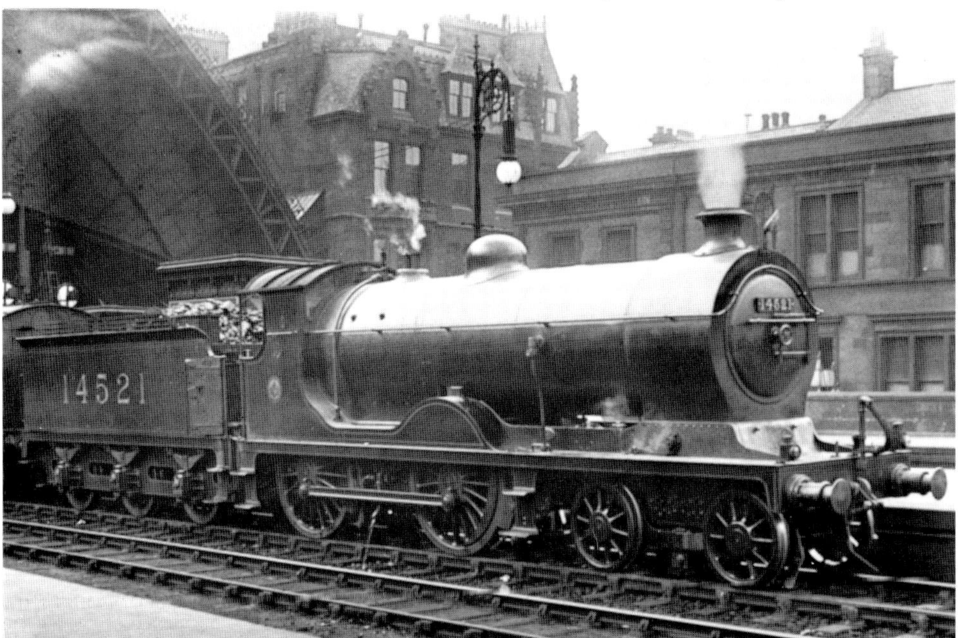

An ex-Caledonian superheated 4–4–0, No. 14494, designed by Pickersgill seen here in early LMS livery. Two of this class worked passenger services between Carlisle and Stranraer in the late twenties. *Courtesy National Railway Museum*

No. 106, one of the B. Conner's 2–4–0s with 6 ft 2 in. driving wheels again as used on the Portpatrick Railway during the years when the CR provided the motive power. *Courtesy National Railway Museum*

No. 689, a 'Jumbo' 0–6–0 by Dugald Drummond as seen on the Portpatrick line in the last years of the nineteenth century. *Courtesy National Railway Museum*

One of Glasgow and South Western Railway Manson's large boilered 4–4–0s with 6 ft 9½ in. wheels (No. 259) which worked the passenger services of the line during the first twenty years of the line's existence seen here in immaculate condition.

Courtesy National Railway Museum

A Class '2P' 4–4–0 No. 463 (Derby built) of the LMS, which were widely used on the Dumfries to Stranraer route during the 20s and 30s.

Courtesy National Railway Museum

A Smellie 'Big Bogie' 4–4–0, No. 14153 in LMS days (re-boilered by Manson). These engines were extensively used on the Portpatrick line.

Courtesy National Railway Museum

crank splashers. The earliest had very primitive front and top protection from the weather for the footplate crew, but later on more substantial cabs were provided, which looked like up-ended coal shovels without handles. These were phased in gradually, and worked the best trains. A small CR 0–4–0 tank engine worked the boat trains on to or off the pier; this structure was not considered stout enough to take anything heavier.

The 'Antiques Road Show' assortment of locomotives acquired from one source or another by Thomas Wheatley for service on the Wigtownshire Railway began with No. 1, a 2–2–2 well tank originally built for the NBR in 1856. The photograph of it on page 41 shows it after heading a special one-coach inspection train from Newton Stewart to Wigtown, together with footplate and station staff, Wheatley himself standing third from the right. A 'single driver' was not the ideal type for this route, and Wheatley rebuilt it as a 2–4–0 in 1883 – one of the last things he did, for he died the same year. The frames had to be altered, and a new pair of driving wheels were supplied from Cowlairs for coupling to the existing single-driven pair; all this was done successfully at Wigtown despite its exiguous facilities. The second locomotive was even older, being a double-framed 0–4–2 first built in 1848 for the Edinburgh and Northern Railway; after absorbing this line the NBR rebuilt the engine in 1870 as a pannier tank. Wheatley bought it in the summer of 1875 and it became No. 2.

No. 3 was also a 0–4–2 tank which was only 15 years old when Wheatley purchased it from the Lancashire and Yorkshire Railway in May 1876; it bore a name, *Addison*, as also did No. 4, *Gardner*, obtained from the same railway; both retained their names. No. 3 had been built by Sharp Stewart and Co., No. 4 by Beyer Peacock & Co. Both worked on the Fleetwood, Preston and West Riding Junction Railway before being sold to Wheatley. No. 4 suffered two rebuildings when in Wigtownshire Railway service, being first given a tender and then being re-converted into a tank engine. No. 5 was another second-hand well tank 2–2–2, bought from the NBR in 1877 and similar to No. 1 in its original state; the Wigtownshire acquired it in 1877, and it was rebuilt in 1885 as a 6-coupled engine with very small wheels by Messrs Shanks and Sons at Arbroath.

No. 6 was a 0–6–0 saddle tank, built by Manning Wardle and Co. in 1866, which had quite a varied career. It was first owned by Brassey and Co., the railway contractors, who used it to construct the Solway Junction line which linked Annan with North Cumberland across the Solway Viaduct (a structure long since dismantled). The Solway Junction Railway then took it over, and when the CR began working this line in 1869 it passed into the latter's locomotive stock. Wheatley eventually bought it, but at first used it as a means of increasing his income by hiring it out to other contractors. At some point in its career it acquired the undistinguished name of *Bradby*, which it kept to the end. At last, in 1882, in a seedy and run-down condition, it was brought back to the Wigtownshire Railway to operate the 1¼-mile branch line between Millisle and Garlieston. Wheatley tidied it up and fitted it with a more capacious cab, and it puffed up and down the Garlieston branch until 1894. No one knows what eventually happened to it.

A first class sleeping carriage such as used on the night service from St Pancras to Stranraer during the 1900s. *Courtesy National Railway Museum*

This Pullman sleeping car was used on the St Pancras to Stranraer service during the late nineteenth century. Note the central side door and ornate decoration more appropriate to the Victorian era. *Author's Collection*

Interior of a 1st class sleeping compartment as used on the MR through night service to Stranraer from London in the 1900s. *Courtesy National Railway Museum*

All the Wigtownshire Railway locomotives except No. 3 were painted green; the latter made up for its black paint by being lined out in red and having its chimney cap and splasher rims edged with brass. Each locomotive bore a plaque which was marked: T. WHEATLEY, OWNER.

After the agreement of 1885 by which the Portpatrick and Wigtownshire Joint Committee was set up, the CR and G&SWR, stifling their mutual dislike, shared the working of the Joint Board's trains between them and alternated in controlling their use. The Lockerbie route, always inconvenient to the passenger proceeding to, from or through Carlisle, was now no longer used for through trains to and from England, since the Annan route was shorter and required no reversals. Glasgow and South Western engines now began to haul most of the passenger trains, particularly those connecting with the boats to Larne. For several years one of Hugh Smellie's '57 class' of domeless 2–4–0s worked the early through train from Carlisle to Stranraer and brought it back in the evening. This was No. 18, a handsome and sturdy machine which fitted comfortably on to the Stranraer turntable. For some unknown reason it was known as 'The Eagle'. By the beginning of the 20th century it was growing old, and the small turntable had been replaced by a somewhat larger one able to take a 4–4–0, so 'The Eagle' was made the pilot engine at Stranraer.

Two of Smellie's 'Wee Bogies', Nos. 87 and 88, took its place. These were 4–4–0s built at Kilmarnock between 1882 and 1885, and were more powerful than No. 18, though the heating surface and grate area in each were the same and the steam pressure was unchanged at 140 lb. per square inch; the cylinders were slightly greater in diameter, however, and the driving wheels 7 in. less wide. The 'Wee Bogies' were very handsome engines, greatly resembling those aristocratic dandies, the SE&CR 'D' class 4–4–0s, though not so decoratively liveried. At the time when they came out they were the most powerful passenger locomotives anywhere in Great Britain. No. 88, unfortunately, was accident-prone, and eventually was so badly damaged by running into a fall of rock near the Big Fleet viaduct that it had to be replaced.

Meanwhile, on the freight services the CR Drummond 'Jumbo' 0–6–0 predominated. This type was one of Drummond's great successes, simple, sturdy and reliable, able to tackle all kinds of duties with success except the haulage of express trains. Drummond built 83 between 1883 and 1887 and his immediate successors built almost as many more; after rebuilding many lasted into LMS days and some even survived to be taken over by BR. Drummond had served at Brighton as Stroudley's assistant, and the resemblance of the 'Jumbos' to the latter's 'C' class was very close, even down to the oval number plate on the cab side. Only the dome, which had Ramsbottom safety valves in place of the Salter type, and the chimney, which was not quite so tall and lacked the splendour of a copper cap, were noticeably different. The tenders, too, were almost identical.

James Manson became locomotive superintendent at Kilmarnock in 1890, and during his period in office, until 1912, he constructed a number of 4–4–0 types, from one of which, the '240' class, built during 1904–1906, some were chosen to take over the boat train working between Carlisle and Stranraer and remain in charge for nerly 10 years. They were larger and more forceful than their predecessors, increased boiler pressure more than compensating for an increase in driving wheel diameter.

Towards the end of the line's independent existence a few of Dugald Drummond's original batch of 4–4–0 express engines, built in 1884, were put to work on the Joint Committee's passenger services; they were now in their old age but had all been rebuilt with new boilers. In their youth they had been Drummond's great success story, enabling some acceleration of trains, a great reduction in piloting and the production of record-breaking runs during the Race to Aberdeen in August 1895.

In September 1915 one of Peter Drummond's new superheated 4–4–0s No. 152, was allotted to Stranraer to run the boat train to Carlisle and back. When it was first put on this service it seemed too good to be true; the job was being done night after night on a coal consumption only half that of its Manson precursor. But it was heavy, and the breaking of some rail-chairs enforced its removal. It was now wartime, and track maintenance was getting into arrears, inevitably when young men were having to go into the Army instead of into civilian employment. This was part of the trouble; the other part lay in the locomotive itself, which for its size was overweight.

These were the principal types of engine which worked between Dumfries and Stranraer during the six decades of the line's independent existence. It is not pretended that these were the only ones, but to give a comprehensive list would be very difficult. In 1923 came the Grouping, when everything changed. The LMS company had at its disposal locomotives from all the groups it had absorbed so that, from 1923 onwards, quite a number of engines from all its constituent companies except the LNWR could at one time or another be seen between Castle Douglas and Portpatrick or on the Whithorn branch. There was a tendency for former CR and G&SWR engines to disappear and give place to newcomers. The CR 0–6–0 'Jumbos' almost vanished from the scene, but in April 1927 other CR engines put in an appearance, notably two ex-CR Pickersgill 4–4–0s Nos. 14492 and 14493, which had been equipped with superheaters. These engines had a reputation for sluggish downhill running, but this did not matter on a line where speed was not really possible, and they were good at uphill work.

After the LMS had built a large number of 3-cylinder 4–4–0 compounds of the former S.W. Johnson MR pattern, as modified by Deeley, bringing the total number up to 240, these were sent to all parts of the system and some came to Galloway; mostly they were used on the Glasgow–Ayr–Stranraer services but a few were rostered to work from Carlisle. They were well liked once the drivers got used to the technique of driving a compound. The ones that worked into and out of Stranraer were from the last-built batch, which had cut-down chimneys and domes and driving wheels of 6 ft 9 in. instead of 7 ft. Unfortunately no records exist to show how well they performed.

In 1929 class '2P' 4–4–0s, recently built at Derby, were sent to the line, and for a while handled most of the passenger trains along with the already-mentioned Pickersgill engines. The class '2P' was a development of a locomotive first put on the Midland main line in the early years of the century, designed by S.W. Johnson and improved by Deeley and Fowler, with superheating and driving wheels of slightly smaller diameter. By now, however, the increasingly heavy early morning and late evening boat trains, whose loads had mounted since third class sleeping cars had been added to their formation, required regular piloting, and the question arose, could larger engines be drafted on to this service? The limiting factor was Stranraer's short turntable, which precluded the use of 4–6–0s.

However, during the short period of his superintendency, G. Hughes, formerly locomotive superintendent at Horwich on the L&YR, had built some 2–6–0s for mixed traffic work. They were peculiar-looking objects by contemporary standards of taste, with two large tilted outside cylinders, outside valve gear and running boards raised high above the cylinders and motion for half the distance between cab and buffer-beam. For some reason they were known as 'Crabs'. Some were sent to Stranraer, and they proved very successful, the gradients of the lines to Carlisle and Ayr holding no terrors for them. These were the first six-coupled engines to be used for passenger train haulage in this part of Scotland, and they just managed to fit on to the Stranraer turntable. They predominated during the 1930s, until loads began to exceed even their abilities, especially on the line to Girvan. Consequently, early in 1939, Stranraer was given a new 60 ft turntable, and 4–6–0s could be used at last.

The engine for the job was also now available – Stanier's celebrated 'Black Five' 2-cylinder 4–6–0, of which over 800 were built and which were to be found all over the LMS system. They could tackle almost any assignment they were given, from fast expresses to slow freight trains, and frequently exceeded 80 mph despite the fact that their wheels were only 6 ft in diameter. They came only just in time, for once World War II had begun the Dumfries–Stranraer line became very busy with wartime traffic. They proved great favourites with the engine crews, and were the mainstay of the old Portpatrick line right to the end of its existence.

To them were added, during the War years, a number of Stanier's other 4–6–0 design, the 'Jubilees', which often worked through from south of the Border. They had by this time overcome the teething troubles which had beset them earlier, but there was much less use for them in England now that the running of fast light expresses had been discontinued. Like the 'Black Fives' they coped well with the heavy trains and were interchangeable with them. For the first time named locomotives were seen on this line, if one excepts the two Wigtownshire Railway tank engines, which had now long been scrapped. They continued in use when the War was over, once they were superseded on the fast light trains they had been built to operate by BR standard 'Britannias'. David L. Smith, in his *Little Railways of South West Scotland*, describes an occasion in August 1953 when the up evening boat train, with a great load of about 450 tons, had a 'Jubilee', No. 45657, *Tyrwhitt*, as train engine, together with a '2P' as pilot, on a night of rain and poor visibility. Three miles short of Newton Stewart the pilot engine failed completely, having a burst-open cylinder cover. After a long delay *Tyrwhitt* was able to push the damaged engine as far as Newton Stewart, where it was detached, and had then to face the long bank from Palnure to Gatehouse of

Fleet, 7 miles of almost unbroken 1 in 80, with a train over 100 tons over its rated maximum. By dint of charging it through going all-out on the preceding 3 miles to the Cree viaduct, the climb was successfully managed, quantities of sand being deposited to prevent the wheels from irretrievably slipping on the wet rails, and the summit was topped; so much water had been used, however, that it was necessary to stop at Loch Skerrow to refill the tank. Here another pilot engine was waiting to assist, but as it had itself been damaged by the collapse of the brick arch in its firebox it could be of little use, so the driver of *Tyrwhitt* went on alone, and reached Dumfries 3½ hours late when it was almost dawn. Here another engine took over and *Tyrwhitt* was turned, to take the down boat train to Stranraer; since this was also late the engine crew had time for a little rest and refreshment.

After the absorption of the LMS into British Railways and the production of the Riddles standard types, there were occasional visits to Stranraer of class '6' 'Clan' Pacifics working through from Carlisle. These engines, impressive to look at, were never as good as their appearance suggested, and only 10 were built in consequence. However, on one occasion a 'Clan' made a record run between Dumfries and Stranraer, when No. 72005, *Clan Macgregor*, with a featherweight load of only three corridor coaches, ran from Dumfries to Stranraer Harbour, 73.8 miles, in 88 minutes 18 seconds actual travelling time, with four intermediate stops, thus picking up 25 minutes of a late start. The fastest section was from Newton Stewart to Stranraer, where a maximum of 72 mph was attained near Dunragit. Had it not been necessary to slack several times in order to set down and pick up tablets the run could probably have been done in 80 minutes actual travelling time. While with such a small load this performance cannot really be regarded as a feat of excellence, it was a creditable effort, and almost certainly no faster run on this line was ever made by any other engine.

To turn now to the coaching stock used on the line: there is perhaps not so much to be said. Old photographs show that purely local trains were composed of 4- or 6-wheel stock until the beginning of the present century, after which cascaded bogie coaches gradually replaced them, at first from other parts of the CR and G&SWR, and after the groupings from almost any part of the LMS. Vestibuled coaches would of course also appear on through trains from the south once the use of corridor stock became general. Interest in the coaches used centres chiefly on the different kinds of sleeping carriage.

The earliest of these were MR Pullman sleeping cars on through trains from and to St Pancras, London. When the Portpatrick and Wigtownshire Joint Committee was formed in 1885 one of the first consequences was the running of through passenger vehicles by the Midland route to Carlisle, which included a recently-built 1st Class sleeping Pullman car of American origin and assembled at Derby. Their advent was hailed by the *Wigtownshire Free Press* of 8th October, 1885, as 'a model of modern art'. It was not a large vehicle – only 36 ft 3 in. long – though it ran on two bogies. Entry was by a centrally placed door on either side; one then turned either right or left to where two large compartments were separated by a toilet section. Each room was so arranged that by day wide seats on either side of the central gangway faced each other across tables; by night the tables were folded away and mattresses were placed over the seat-bases so that each seat back became one end of a bed arranged longitudinally between them. At the same time corresponding upper berths, folded away in the daytime against the curved side of the roof on either side of the clerestory, were let down above the lower berths. In this manner 16 persons could be accommodated. There were no end vestibule connections.

The newspaper praised the interior decoration. The rooms were 'beautifully carved and finished', lined with 'mahogany inlaid with white maple, boxwood and ebony'. Each seat was sprung and 'covered with flowered claret plush . . . Each berth is richly decorated with flowered curtains made of silk and wool . . . The whole carriage is heated with hot water pipes from one of Baker's patent American stoves'. One of these sumptuous vehicles ran nightly between St Pancras and Stranraer on each of the boat trains. They were in use for six years; after that they were taken out of service and no through sleeping accommodation was provided until 1904, when, in the words of the G&SWR timetable, 'new and luxurious sleeping carriages' were again brought into service on this route.

The new vehicles were built in the more familiar pattern of the present day, with side corridors from which six single compartments and two double compartments led off; in addition there was a 'smoking room' at one end and an attendant's room at the other, and a toilet compartment at either end. Each single compartment had a single bed and a washbasin; each double compartment,two beds and a washbasin. Heating was by steam pipes in which the flow could be regulated by the passenger. Electric bells were installed for summoning the attendant, who supplied light refreshments on request. All lighting was by electricity. A fire extinguisher was provided in case of emergency. The coach ran on six-wheel bogies to ensure smooth riding. Altogether it was a considerable improvement on the previous much-praised Pullman vehicles.

At the same time the timetable advertised some provision of overnight comfort for the third class traveller. Rugs and pillows could be hired out at the start of the journey for 6*d*. each, so that if the train were fairly empty their hirer could make himself physically almost as comfortable as his first class cousin by stretching himself out along one side-cushion – though of course he could not guarantee he would have its exclusive use for the whole journey. This provision, and the inclusion of first class sleepers, continued to be advertised until 1914, after which nothing was said in the timetable about them, though first class sleepers did continue to run for the benefit of travellers going to or coming from Northern Ireland, subject to a maximum of 10 available places.

After the war the through sleeping services returned, but now from and to Euston, using West Coast Joint Stock sleeping cars. In 1931 third class sleepers were added to the train; these offered what would now be known as 'couchette' accommodation, with four bunks in each compartment, two above and two below, and the provision of rugs and pillows; toilet and washing facilities were found at either end. If necessary an attendant could be summoned from the adjoining first class vehicle. Right at the end, after the formation of British Railways, the new type of second class sleeping car replaced the 'couchette' variety, with full bedding facilities and a small wash-hand basin in each two-berth compartment.

No restaurant or buffet cars were ever included in the boat train formations. When there was a through daytime service one could dine in a restaurant car before the Stranraer portion was detached at Carlisle, and on the night boat train from Euston a dining car was attached at the rear, which was removed at Crewe, since the train left at 7.30 pm and such provision was obviously needed. The writer has a vivid memory of a journey made to Newton Stewart from London in August 1953, when, having failed to get a 'couchette' berth since the train was crowded, he spent the money on an evening meal instead. Sitting in the restaurant car at the end of the train, he awaited being served; opposite him was an elderly lady of severe aspect wearing a stylish costume. The meal did not begin until we had been under way for an hour; then, as we waited to pick up passengers at Bletchley, the attendant came round with the soup tureen and proceeded to ladle its contents on to the plates. He managed to fill my plate successfully, but as he was doing the same for the lady opposite the train gave a great lurch as it started and the contents of the ladle went all down her costume. In considerable and understandable annoyance she rose and went back to her compartment – presumably a sleeper where she could change her clothes. The attendant was very distressed; evidently he expected to be blamed, though it was in no way his fault. In the days of steam this sort of thing often happened (though not necessarily with the same unhappy consequence) if a driver had to set back against the springs of the train's buffers, because he had stopped with the driving crank at front dead centre (and hence the piston was at the end of its stroke).

The first BR Standard Class '5', 4–6–0, No. 73000. A few of this class worked between Carlisle and Stranraer during the 1950s. *Courtesy National Railway Museum*

Chapter Fourteen

Passenger Train Services

When the Portpatrick Railway first began to operate the train service was slow by later standards, though very much faster than anything horse-drawn vehicles could achieve. The 53¼ miles from Castle Douglas to Stranraer took 2½ hours by the morning train and ten minutes more by the evening train; in the opposite direction the times were 2 hours 32 minutes and 2 hours 40 minutes respectively, all seven intermediate stops being made. The morning down train had made connections with overnight trains from the south, but none are shown on the timetable for the down evening train, which did, however, have a connection from Glasgow *via* Dumfries. The early morning up service from Stranraer also shows a Glasgow connection, and both it and the afternoon train had connections for London. In either direction there were long waits at Carlisle.

The following year, 1862, the first trains ran between Stranraer and Portpatrick, two each way a day, taking between 21 and 27 minutes with a stop at Colfin. By now, too, there was an additional train each way between Castle Douglas and Stranraer. This was the best the Portpatrick Railway could offer. If one were going to or coming from south of the Border, one changed at Castle Douglas, Dumfries and Carlisle.

However, once the CR took over the working of the line in 1864 frequencies increased, though not greatly. An extra train to meet local needs ran on Fridays only from Glenluce to Stranraer. Four up trains left the latter place daily for Castle Douglas, though there were still only three down trains. The Stranraer–Portpatrick service gained an extra train a day. The line between Dumfries and Lockerbie was now in use, and through trains to Carlisle were routed that way. In 1867, although extra stops had to be made at the newly-built stations at Castle Kennedy, Dunragit, Palnure and Dromore (later Gatehouse of Fleet), timings were quicker – as little as 2 hours 15 minutes from Castle Douglas to Stranraer by the down morning train and 2 hours 18 minutes in the opposite direction by the evening train.

The 1870 timetable shows four down trains covering the whole distance; the second (which connected with the boat to Larne) took only 1¾ hrs from Castle Douglas to Stranraer Harbour; the corresponding up evening boat train was similarly timed. A local third class-only train now made an appearance, running in the early afternoon from Newton Stewart to Stranraer; this was actually a goods train with coaches attached, and because of shunting operations performed *en route* it required 100 minutes to do the 23½ mile journey! It was known as the 'wee train'. In the up direction as many as five trains are shown, running the whole length of the line, the last having connected with the steamer from Larne. This was the best daily service the line was to have for some while. By 1873 there were only four trains each way, with the 'wee train' now returning to Newton Stewart in the late afternoon, again on a very slow schedule. The 1873 pattern continued to 1885, when the agreement between the Portpatrick Railway and the CR ran out, the only change being that the 'wee train' now left attached to an early morning goods train from Newton Stewart instead of an afternoon one, so giving much more time for shopping expeditions in Stranraer.

DUMFRIES to CASTLE-DOUGLAS, KIRKCUDBRIGHT, and STRANRAER.

St. Enoch Station.	p.m	p.m	..	a.m	a.m	a.m	a.m	a.m	a.m	..	..	n'n	..	p.m	p.m	p.m	p.m	p.m	..	p.m
36 Glasgow .... lev. b	9†15	9§15	†..	..	..	..	..	..	9 20	..	..	12 0	..	1212	2 0	2 0	5 10	5 30	..	1110*
36 Kilmarnock ″ b	9†52	9§52	†..	..	..	5 55	5 55	..	9 57	..	..	1239	..	1252	2 41	2 41	6 10	6 10	..	1141*
50 Ayr .... ″ b	9†0	9§0	†..	..	..	..	..	..	9 3	..	..	1140	..	1110	2ƒ18	2ƒ18	5 20	5 20	..	1 55*
38 Carlisle .... ″	2a10	3a13	..	4 45		7 0	7 0	..	1015	..	..	1p26		1p25	4 40	4 40	6 47	6 47	..	6a20
Dumfries .... lev.	3 30	4 5	..	6 30	Lev. Whithorn 7-35 a.m. and Wigtown 8-5 a.m.	8 35	9 0	..	1130	..	..	2 38	..	3 12	5 30	5 37	7 37	7 45	..	7 15
Maxwelltown	a.m	a.m	..	6 36		..	9 6	..	1136	..	..	2 44	..	3 18	..	5 43	..	7 51	..	a.m
Lochanhead	..	..	..	6 45		..	9 15	..	1145	..	..	2 53	..	3 27	..	5 52	..	8 0	..	..
Killywhan	..	..	..	6 50		..	9 20	..	1150	..	..	2 58	..	3 32	..	5 57	..	8 5	..	..
Kirkgunzeon	..	..	..	6 55		..	9 26	..	1156	..	..	3 4	..	3 38	..	6 3	..	8 11	..	..
Southwick	..	..	..	7 2		..	9 32	..	12 2	..	..	3 10	..	3 44	..	6 9	..	8 17	..	..
Dalbeattie	..	4 [illegible]	..	7 10		9 [illegible]	9 39	..	12 8	..	..	3 16	..	3 51	..	6 14	8 2	8 22	..	7 47
Castle-Douglas .. arr.	4 3	4 40	..	7 23		9 13	9 50		1218	..	..	3 28	..	4 0	6 3	6 21	8 12	8 32	..	8 0
Castle-Douglas lev.	..	4 45	..	7 25		..	1010	1145	1220	..	..	3†40	..	4 2	..	6 27	..	8 35	..	☞ Sundays only.
Bridge of Dee	..	..	..	7 32		..	1017	1152	1227	..	..	3†47	..	4 9	..	6 34	..	8 42	..	
Tarff	..	5 2	..	7 43		..	1027	12 2	1237	..	..	3†57	..	4 18	..	6 44	..	8 52	..	
Kirkcudbright arr.	..	5 10	..	7 53	☞	..	1035	1210	1245	..	..	4†5	..	4 25	..	6 52	..	9 0	..	
Castle-Douglas lev.	4 5	4 42	..	..		9 17	..	..	..	..	..	3 35	..	☞ Saturdays only.	6 5	..	8 13	..	..	
New Galloway	..	..	..	..		9 37	..	..	..	..	..	3 54	..		6 22	..	8 31	..	..	
Newton-Stewart	4 58	5 35	..	..	8 36	1024	..	..	..	..	..	4 42	..		7 0	..	9 15	..	..	
Dunragit	..	..	..	..	9 13	1111	..	..	..	..	..	5 28	..		..	..	9 54	..	..	
Castle-Kennedy	..	..	..	..	9 20	1118	..	..	..	..	..	5 35	..		..	..	10 3	..	..	
Stranraer Stn arr.	5h42	6h19	..	..	9 28	1125	..	..	..	..	..	5 43	..		7h44	..	1013h	..	..	..
44 Dunragit .... lev.	..	..	..	..	..	1150d	..	..	..	..	..	..	..		..	..	..	..	..	..
44 Girvan .... arr.	8c32	..	..	..	..	1d4	..	..	..	..	..	..	..		D	..	..	..	..	..
44 Ayr .... ″	9c15	..	..	..	..	1d47	..	..	..	..	..	..	..		11‖5	..	..	..	..	..
44 Glasgow (St En) ″	1028c	c..	..	..	..	3d2	..	..	..	..	..	..	..		12‖5	..	..	..	..	..

See Page 57.

* Saturday nights † Except Saturdays § Mondays only h Harbour Station ƒ via Mauchline
b On Sundays leaves Glasgow at 9-15 p.m, Kilmarnock at 9 55 p.m, and Ayr 9 0 p.m, arr. Dumfries 11-14 p.m
c *Via* Stranraer, leaving Town Station at 7-0 a.m, and pay excess fare as from Dunragit to Stranraer.
d Connection from Portpatrick and Wigtownshire Lines formed by changing at Dunragit.
‖ *Via* Stranraer, leave 9-25 p.m. and pay excess fare from Dunragit D Sets down passengers off Irish Steamer only

STRANRAER, KIRKCUDBRIGHT, and CASTLE-DOUGLAS to DUMFRIES

St. Enoch Station.	a.m	a.m	a.m	..	a.m	p.m	p.m	p.m	p.m	..	p.m	p.m	p.m	p.m	..		p.m	..	..	p.m	..
40 Glasgow ...... lev.	..	..	..	..	5b0E	Saturdays only.	Except Saturdays	..	..	..	..b	1230	..	4b10	..	Departure contingent on arrival of Irish Steamer at Stranraer.	4b10	..	..	Sundays only. ☞	..
40 Ayr .... ″	..	..	..	..	7b5			..	..	..	..	1b30	..	5b20	..		5b20	..	..		..
40 Girvan .... ″	..	..	..	..	7b50			..	..	..	..	2b13	..	6b3	..		6b3	..	..		..
40 Dunragit ...... arr.	..	..	..	..	9b0			..	..	..	..	3b26	..	7b15	..		..	..	..		..
Stranraer Stn. lev.	..	..	7 35	..	9 30			1215h	..	..	..	3 40	..	6 55	..		9h42	..	..		..
Castle Kennedy	..	..	7 42	..	9 37			..	..	..	..	3 48	..	7 3	..		..	..	..		..
Dunragit	..	..	7 51	..	9 45			..	..	..	..	4 0	..	7 17	..		..	..	..		..
Newton-Stewart. ..	..	..	8 42	..	1031			1 3	..	..	..	4 41	..	7 55	..		1030	..	..		..
New Galloway, ...	..	..	9 3[illegible]	..	1115	..	..	1 45	..	..	..	5 34	..	Wigtown arrive 8.15, and Whithorn 8-50 p.m. ☞	..		..	..	..		..
Castle-Douglas..arr.	..	..	10 0	..	1136	..	..	1 56	..	..	..	5 58	..		..		1123	..	..		..
Kirkcudbright lev.	6 30	8 45	..	..	11 0	1 0	1 0	..	2 45	..	4 45	..	7 15		..		..	..	..		..
Tarff	6 39	8 54	..	..	11 9	1 9	1 9	..	2 54	..	4 57	..	7 24		..		..	..	..		..
Bridge of Dee	6 49	9 4	..	..	1119	1 19	1 19	..	3 4	..	5 7	..	7 34		..		..	..	..		..
Castle-Douglas..arr.	6 54	9 9	..	..	1124	1 24	1 24	..	3 9	..	5 12	..	7 39		..		..	..	..		..
Castle-Douglas .. lev.	7 0	9 15	10 2	..	1138	1 26	..	2 0	3†10	..	5 14	6 0	7 41		..		1125	..	..	9 0	..
Dalbeattie..........	7 12	9 [illegible]	1015	..	1150	1 36	..	2 10	3†30	..	5 24	6 11	7 51		..		..	..	..	9 20	..
Southwick	7 17	9 30	..	..	..	1 4[illegible]	..	..	3†36	..	5 29	..	7 56		..		..	..	..	..	..
Kirkgunzeon........	7 23	9 36	..	..	..	1 47	..	..	3†42	..	5 35	..	8 2		..		..	..	..	..	..
Killywhan..........	7 29	9 42	A	..	..	1 5[illegible]	..	..	3†48	..	5 43	..	8 10		..		..	..	..	..	..
Lochanhead	7 35	9 48	..	..	..	1 59	..	..	3†54	..	5 49	..	8 16		..		..	..	..	..	..
Maxwelltown	7 44	9 57	..	..	..	2 8	..	..	4†4	..	5 58	..	8 24		..		..	..	..	..	..
Dumfries .... arr.	7 50	10 3	1030	..	1215	2 15	..	2 31	4†11		6 5	6 40	8 3[illegible]		..		1156	..	..	9 55	..
36 Carlisle ...... arr.	9 15	1130	12 [illegible]	..	2 40	5 25	..	5 25	6 5	..	8 18	8 18	10 0		..		1242a	..	..	12 6a	..
38 Kilmarnock... ″	1018	1 32	1 [illegible]	..	3 2	4 40	..	[illegible]	6 28	..	8 47	8 47	..		..		k5a42ℓ	..	..	5a42k	..
51 Ayr.......... ″	10[illegible]ƒ	1ƒ20	1ƒ[illegible]	..	3 50	5 45	..	5 45	7 25	..	9 38	9 38	..	..	..		k6a25ℓ	..	..	6a25k	..
38 Glasgow St. En ″	11 5	2 20	2 [illegible]	..	3 43	5 28		5 23	7 11		9 30	9 30	..	..	..		k7a35ℓ	..	..	7a35k	..

See Page 57.

* Sats. only ℓ On Sun. mornings lev. Dumfries 7 6 a.m: arriving Kilmarnock 8-23, Ayr 9-15 and Glasgow 9-7 a.m
E Mondays, Except 16th, leave 5-55 a.m.
A Calls at 10-25 a.m. to set down from beyond Castle Douglas. b Connections formed by changing at Dunragit.
ƒ *Via* Mauchline h Leave Harbour Station ℓ Monday 10 mins. later k Leave Dumfries for Kilmarnock and Ayr at 4-10 a.m. and for Glasgow at 5-31 a.m

DUMFRIES and MONIAIVE (3rd Class only)

St. Enoch Station.	a.m	..	..	n'n	..	..	p.m	..
36 Glasgow ...... lev.	..	..	..	Wednesdays only 12 0	..	..	2 0	..
Dumfries........ lev.	8 35	..	..	2p15	..	..	5 45	..
Irongray	8 50	..	..	2 30	..	..	6 0	..
Newtonairds........	8 58	..	..	2 38	..	..	6 8	..
Stepford	9 4	..	..	2 44	..	..	6 14	..
Dunscore	9 10	..	..	2 50	..	..	6 20	..
Crossford..........	9 20	..	..	3 0	..	..	6 30	..
Kirkland.	9 27	..	..	3 7	..	..	6 37	..
Moniaive .... arr.	9 35	..	..	3 15	..	..	6 45	..

	a.m	..	a.m	..	..	p.m	..
Moniaive ...... lev.	7 0	..	Wednesdays only 9 55	..	..	3 45	..
Kirkland	7 8	..	10 3	..	..	3 53	..
Crossford	7 15	..	1010	..	..	4 0	..
Dunscore	7 25	..	1020	..	..	4 10	..
Stepford.	7 30	..	1026	..	..	4 15	..
Newtonairds	7 35	..	1030	..	..	4 20	..
Irongray	7 45	..	1040	..	..	4 30	..
Dumfries .... arr.	8 0	..	1055	..	..	4 45	..
38 Glasgow St En arr.	11 5	..	2 20	..	..	7 11	..

DUMFRIES, CASTLE-DOUGLAS, NEWTON-STEWART, & STRANRAER.

St. Enoch Station.	p.m	p.m	a.m	a.m	n'n	p.m	p.m
36 Glasgow .... lev. b	9†15	9§15	b..	..	12 0	2 0	5 30
70 ″ (Central) ″	..	..	Leave Whithorn 7-35 a.m. and Wigtown 8-5 a.m.	..	E1025	1ƒ40E	4E10
70 Edinburgh (PrSt) ″	a.m	a.m		..	E1145	1n30E	4E10
38 Carlisle (S.W.) ″	2 40	3 13		7 0	1p26	4 40	6 47
56 Dumfries...... ″	3 30	4 5		8 35	2 38	5 30	7 37
	a.m	a.m		a.m	p.m	p.m.	p.m.
Castle-Douglas .. lev.	4 5	4 42		9 17	Thro' Car Edinburgh (Pr. St.) to Whithorn 3 35	6 5	8 13
Crossmichael	..	..		9 24	3 42	..	8 20
Parton	..	..		9 30	3 48	..	8 26
New Galloway	..	..		9 37	3 54	6 22	8 32
Gatehouse on Fleet ..	..	..		10 0	4 17	..	8 53
Creetown	..	..		1012	4 26	..	9 3
Palnure............	..	..		1018	4 35	..	9 9
Newton-Stewart arr.	4 58	5 35		1024	4 42	7 0	9 15
Newton-Stewart lev.	4 59	5 38	8 36	1031	4 52	7 3	9 18
Kirkcowan	..	..	8 48	1043	5 4	..	9 30
Glenluce	..	..	9 5	11 0	5 20	..	9 45
Dunragit	..	..	9 13	1111	5 28	..	9 54
Castle-Kennedy	..	..	9 20	1118	5 35	..	10 3
Stranraer Station arr.	..	..	9 28	1125	5 43	..	h
Stranraer Harbour ″	5 42	6 19	..	..	..	7 44	1013
Stranraer Boat lev.	6a30	6a30	..	..	..	7 55	6a30
Larne (Harbour) arr.	8 45	8 45	..	..	..	1010	[illegible] 45
Belfast.... Train ″	9 30	9 30	..	..	..	11 5	9 30

	a.m	a.m	a.m	p.m	p.m	p.m	..
Belfast Train .. lev.	..	..	9 5	..	..	6 25	..
Larne (Harbour). ″	..	..	9 55	..	..	7 18	..
Stranraer Harb. arr.	..	..	12 3	..	..	8 20	..
Stranraer Harbor lev.	..	Thro' Car from Whithorn to Edinburgh (Pr. St.) ..	1215	..	..	9 42	..
Stranraer Station ″	7 35	9 30	..	3 40	6 55	..	..
Castle-Kennedy	7 42	9 37	..	3 48	7 3	..	
Dunragit	7 51	9 45	..	4 0	7 17	..	
Glenluce	8 2	9 56	..	4 9	7 28	..	
Kirkcowan	8 18	1011	..	4 25	7 44	..	
Newton-Stewart arr.	8 32	1021	1 0	4 36	7 55	1027	
Newton-Stewart lev.	8 42	1031	1 3	4 46	Arrive Wigtown 8-15 p.m and Whithorn 8-50 p.m	1030	E *Via* Carstairs and Lockerbie.
Palnure..............	8 49	1038	..	4 54		..	
Creetown	8 56	1045	..	5 2		..	
Gatehouse on Fleet ..	9 8	1056	..	5 13		..	
New Galloway	9 37	1115	1 43	5 34		..	
Parton	9 43	1121	..	5 40		..	
Crossmichael	9 50	1128	..	5 48		..	
Castle-Douglas .. arr.	10 0	1136	1 58	5 58		1123	
	a.m	p.m	p.m	p.m.		a.m.	
56 Dumfries ...... arr.	1040	1215	2 34	6 40		1156	
36 Carlisle (S.W.) ″	12 5	2 40	3 25	8 18		1242	
74 Edinburgh (PrSt) ″	3E20	3E20	6E0	11E0		a.m	
74 Glasgow (Cent.) ″	3E20	3E20	6E5	11E0		..	
38 ″ (St. Enoch) ″	2 20	3 43	5 28	9 30		7s35	

† Except Saturdays b Leaves St. Enoch at 9-15 p.m. on Sundays ƒ Sats. 12-45 p.m s Sundays arr 9-7 a.m.
h Passengers for Ireland stay at Stranraer over night, and those holding Cabin Tickets or paying difference between Steerage and Cabin Tickets, plus 4/4½ per berth can sleep on board Steamer. n Sats 12-53 p.m

GLASGOW STRANRAER, NEWTON-STEWART, WIGTOWN, & WHITHORN

St. Enoch Station.	a.m	a.m	a.m.	p.m		p.m	p.m
56 Glasgow via Dunragit lev	..	..	5E0	Fridays only ..	Thro' Car from Edinbro (Pr. St.) lev 11-45 a.m	1230	4 10
56 ″ via Dumfries ″	..	..	..	..		12 0	2 0
38 Carlisle.......... ″	313	313	7 0	..		1 26	4 40
56 Dumfries .... ″	4 5	4 5	8 35	..		2 38	5 30
Newton-Stewart..... lev.	630	835	1050	2 30		4 55	8 0
Wigtown	6 45	8 55	11 5	2 45		5 10	8 15
Kirkinner	650	9 5	1110	2 50		5 16	8 20
Whauphill..........	655	915	1116	2 56		5 22	8 27
Sorbie..............	7 0	925	1122	3 1		5 28	8 33
Millisle (for Garlieston) a	7 4	935	1127	3 6		5 34	8 38
Whithorn .... arr.	715	950	1139	3 15		5 45	8 5[illegible]

	a.m		a.m.	p.m.	p.m	p.m
Whithorn .... lev.	7 35	Thro' Car to Edinburgh (Pr. St. arr 3-20 p.m.	9 25	Fridays only 1210	3 30	6 15
Millisle (for Garlieston) a	7 43		9 33	12 7	3 39	6 22
Sorbie	7 48		9 39	1222	3 45	6 27
Whauphill	7 54		9 47	1227	3 52	6 34
Kirkinner	7 59		9 52	1231	3 57	6 36
Wigtown	8 5		10 1	1235	4 5	6 40
Newton-Stewart..... arr.	8 20		1018	1250	4 20	6 55
56 Dumfries .... arr.	1040		1215	..	6 40	1156
36 Carlisle .... ″	12 5		2 40	..	8 18	1242
56 Glasgow via Dumfries ″	2 20		3 43	..	9 30	7s35a
56 ″ via Dunragit ″	..		3 2	..	..	12d5

† Except Sats. d *Via* Stranraer. h Harbour Station. s Sundays arr. 9-7 a.m. E Mons. Exc. 16th 5 55 a.m
☞ a A Conveyance runs between Millisle and Garlieston in connection with all the Trains.

STRANRAER and PORTPATRICK.

	a.m	a.m	a.m.	p.m	p.m.	..	p.m	..
40 Glasgow (St. En.) lev	..	5E0	9 0	1230	..	..	4 10	..
40 Ayr .... ″	..	7 5	10 3	1 30	..	..	5 10	..
56 Carlisle...... ″	3 13	..	7 0	..	1 26	..	4 40	..
56 Dumfries ...... ″	4 5	..	8 35	..	2 38	..	5 30	
Stranraer Hrbr. lev.	..	..	..	..	..	..	7 50	..
Stranraer Stn.... ″	7 10	9 35	1215	3 50	5 50	..	8 0	..
Colfin..............	7 27	9 45	1225	4 0	6 0	..	8 10	..
Portpatrick ..... arr.	7 40	9 55	1237	4 10	6 10	..	8 20	..

	a.m.	a.m	p.m	p.m	p.m	..	p.m.
Portpatrick...... lev	8 50	1050	2 50	5 0	6 20	..	8 35
Colfin..............	9 3	11 0	3 2	5 10	6 30	..	8 45
Stranraer Stn. .. arr.	9 20	1110	3 20	5 20	6 40	..	8 55
Stranraer Harb.. arr.	..	..	..	..	..	..	9 6
56 Dumfries ...... arr.	1215	..	6 40	..		..	1156
56 Carlisle ...... ″	2 40	..	8 18	..		..	1242a
44 Ayr .... ″	..	1 47	6 17	..	..	..	11 5
44 Glasgow (St. En.) ″	..	3 2	7 39	..		..	12a5

* Sats. only § Mons. only E Mons. Exc. 16th 5-55 a.m.

STRANRAER and DUNRAGIT.

	a.m.	..	a.m	..	a.m	..	a.m	p.m	p.m	..	p.m	..	p.m	..	..	..	..	..	..	..
Stranraer Harbor lev.	..	..	..	..	..	..	..	1225	..	..	..	..	..	..	..	..	..	..	..	..
Stranraer Stn .. ″	7 0	..	7 35	.	9 30	..	1132	..	3 40	..	4 10	..	6 55	..	..	..	..	..	..	..
Castle Kennedy	7 7	..	7 42	..	9 37	..	1139	1234v	3 48	..	4 17	..	7 3		..	..	..	..	..	..
Dunragit .... arr.	7 15	..	7 51	..	9 45	..	1150	1240	4 0	..	4 25	.	7 17	..	..	..	..	..	..	..

	a.m	..	a.m	..	a.m	..	a.m	p.m	p.m	..	p.m	..	p.m.	p.m	..	..	..	..	..	..
Dunragit .... lev.	9 0	..	9 13	..	1111	..	1150	3 26	5 28	..	7 15	..	9 0	9 51	..	..	..	..	..	..
Castle-Kennedy	9 7	..	9 20	..	1118	..	1157	3 33	5 35	..	..	..	9 7	10 3	..	..	..	..	..	..
Stranraer Stn .. arr.	9 15	..	9 28	..	1125	..	12 5	3 33	5 43	..	..	..	..	..	..	..	..	..	..	..
Stranraer Harbour ″	..	..	..	..	..	..	..		..	..	7 28	..	9 15	1013	..	..	..	..	..	..

v Weds. & Sats. only

The Murray's Railway Guide passenger timetable for July 1923.

After the formation of the Joint Committee in 1885 the through boat train services from the South began to take the Annan route of the G&SWR instead of going via Lockerbie, and reached Castle Douglas earlier. In 1891 the down departure from the latter station was 5.21 am; two years later it had become 4.54; from 1897 to 1900 it was 4.20; soon afterwards it stabilised at 4.26 until 1914, with Stranraer Harbour being reached in time for a boat departure at 6 am and a Belfast arrival time of 8.35 am. Throughout this period this train carried the mails, and also through passengers from St Pancras in London over the Midland route to Carlisle; except for a few years after 1890 there was also a through sleeping car service for first class passengers. This train can have carried little local traffic. In the opposite direction increasingly later departures from Stranraer, corresponding with later departures from Belfast, stabilised for a while at 8.50 pm from Stranraer Harbour, and then became later still; from 1906 to 1914 the advertised departure was 10.03 pm (6.30 pm from Belfast).

In regard to the other trains which ran the whole length of the line and called at all stations: during 1890–1914, in the down direction, there was a morning service from Castle Douglas at about 9.30 am (from 1906 stabilising at 9.25), an afternoon one at about 3.15 pm (3.30 pm from 1903), a late evening one at about 8 pm, and an early evening one, omitting several stops, at about 6 pm when there was an evening summer sailing from Stranraer to Larne; in years when no evening sailing was advertised the last train was transferred to the boat train slot and called at all stations. In the up direction departures from Stranraer were less evenly spaced out during the same period; an early train leaving there a little after 8 am gradually became earlier; a mid-morning train leaving at about 10 am did the same; an afternoon train at 3.40 remained at that departure time throughout. From 1906 to 1914 the times stabilised at 7.30 am, 9.30 am and 3.40 pm. By this time, too, the local 'wee train' had settled down to a morning departure from Newton Stewart at 8.32 am, returning from Stranraer a little before or after 7 pm; it connected with trains on the Wigtownshire branch and in some years actually itself ran from or to Whithorn.

The Stranraer–Portpatrick service generally remained at three trains daily each way, early morning, mid-day and late evening outwards and later morning, mid-afternoon and early evening returning; in some years extra trains were added, and in 1911 as many as six in each direction were advertised; times between stations varied between 20 and 25 minutes.

World War I brought about a worsening of services. In 1918 only four trains ran from Castle Douglas, the early morning boat train arriving much later, with no definitely-advertised arrival times at either Larne or Belfast. This was of course because the threat of submarine attack was now compelling the packet steamer to follow a zig-zag course so that the passage time was increased indefinitely. There were only three up trains from Stranraer going the whole way, the mid-morning one terminating at Newton Stewart. The 'wee train' still ran at its usual times. By 1922, however, things had improved. The boat train now ran earlier to and later from Stranraer, and the first class sleeping cars were again advertised (one had always been available during the War years, but with only a limited number of berths). The

Dumfries, Castle-Douglas, Kirkcudbright, Newton-Stewart, and Stranraer

Table 17

Miles	WEEKDAYS ONLY	p.m	a.m	a.m	a.m	Sats only a.m	p.m	p.m	p.m	p.m	..
..	62 EDINBURGH (Wav.) K	7d0†	..	..	..	7 35	11a0J	..	2 0	2 0	..
..	13 GLASGOW (St. Enoch) lev.	9†b5	..	..	..	9 25	1250F	..	4 0	4 0	..
..	13 CARLISLE	2a37	5D41	7 0	7 45	11 8	1 19	..	5 7	5 7	..
..	Dumfries lev.	3 25	6 35	8 10	8 45	1222	2 50	..	6 2	6 30	..
8¼	Killywhan	..	..	8 25	..	..	..	..	6 17	..	..
12½	Southwick	..	..	8 33	..	1243	..	..	6 26	..	..
14¼	Dalbeattie	..	7 3	8 39	9 10	1249	3 14	..	6 30	6 53	..
19¾	Castle-Douglas arr.	3 54	7 12	8 48	9 18	1256	3 23	..	6 39	7 1	..
..	Castle-Douglas lev.	..	7 16	8 51	..	1 10	..	3 30	6 40	..	..
26½	Tarff	..	7 29	9 3	..	..	..	3 41	6 51	..	..
30	Kirkcudbright arr.	..	7 36	9 10	..	1 28	..	3 48	6 58	..	..
..	Castle-Douglas lev.	4 0	..	..	9 21	..	3 25	..	..	7 3	..
23¼	Crossmichael	..	..	..	9 29	..	3 32	..	..	7 10	..
26½	Parton	..	..	..	9 35	..	3 38	..	..	7 17	..
28¾	New Galloway	..	..	..	9 45	..	3 44	..	..	7 23	..
32¾	Lochskerrow	..	..	..	9 57	..	3 56	..	..	7 35	..
38¾	Gatehouse-of-Fleet	..	..	..	10T7	..	4P7	..	..	7P46	..
43¼	Creetown	..	..	..	1019	..	4 18	..	..	7A56	..
49½	Newton-Stewart arr.	4 54	..	..	1031	..	4 30	..	..	8A8	..
..	Newton-Stewart lev.	5 2	..	..	1033	..	4 34	..	..	8A10	..
55¾	Kirkcowan	..	..	..	1045	..	4 46	..	..	..	..
64¼	Glenluce	..	..	..	11 0	..	5 1	..	..	8A35	..
67½	37 Dunragit	..	..	..	1110	..	5 8	..	..	8A43	..
70½	37 Castle-Kennedy	..	..	..	1119	..	5 15	..	..	..	..
73¾	37 Stranraer Harbour arr.	5 45	..	..	..	..	..	..	..	..	..
73	37 Stranraer (Town) arr.	6L37	..	..	1125	..	5 22	..	..	8A53	..
..	Stranraer Harbour lev.	7 0	..	..	..	..	..	..	..	..	..
..	Larne Harbour arr.	9 15	..	..	..	..	..	..	..	..	..
..	Belfast (York Road)	10 5	..	..	..	..	..	..	..	..	..
..	Londonderry arr.	1 30	..	..	..	..	..	..	..	..	..
..	37 DUNRAGIT lev.	M	..	..	1159h	..	..	..	..	..	..
..	37 GLASGOW (St. Enoch) arr.	1021g	..	..	2h56	..	..	..	..	..	..

(First column: The Northern Irishman TC and SC London (Euston) leave B 7-30 p.m to Stranraer Harbour. Passengers may remain in Berths until 6-25 a.m.)

B Larne and Stranraer Steamer Train
B Leave 7-40 p.m. on Fridays from 23rd May, 1958
b Sundays lev. Glasgow (Central) at 9-20 p.m St. Enoch Station from 18th May, 1958
D Mons. 5-29 a.m d Suns. 6-40 p.m
F Saturdays 12-15 p.m.
g Via Stranraer leave Town Station 7-25 a.m.
h Connection formed by changing at Dunragit
H Sundays 10-22 a.m., change Central and Queen St. Stations from 29th September, 1957, until 11th May, 1958, inclusive
J Saturdays 10-0 a.m.

Stranraer, Newton-Stewart, Kirkcudbright, Castle-Douglas and Dumfries

WEEKDAYS ONLY	a.m	a.m	a.m	..	Sats only a.m	Ex Sats a.m	..	Sats only p.m	p.m	p.m	p.m	p.m
37 GLASGOW (St. Enoch) lev.	..	..	..	..	..	..	..	..	..	1230h	..	5K10
37 DUNRAGIT arr.	..	..	..	..	..	..	..	..	..	3h18	..	A
Londonderry lev.	..	..	..	..	..	..	..	..	..	..	..	3 15
Belfast (York Road)	..	..	..	..	..	..	..	..	..	..	..	5 55
Larne Harbour	..	..	..	..	..	..	..	..	..	..	..	6 50
Stranraer Harbour arr.	..	..	..	..	..	..	..	..	..	..	..	8 55
37 Stranraer (Town) lev.	..	..	8 5	..	1058	1058	..	..	..	3 40	..	..
37 Stranraer Harbour	..	..	..	..	..	..	..	..	..	..	..	10 0
37 Castle-Kennedy	..	..	8 11	..	11 4	..	..	..	..	3 46	..	..
37 Dunragit	..	..	8 19	..	1112	1112	..	..	..	3 55	..	..
Glenluce	..	..	8 27	..	1121	1121	..	..	..	4 5	..	..
Kirkcowan	..	..	8 42	..	1136	..	..	..	..	4 21	..	..
Newton-Stewart arr.	..	..	8 53	..	1147	1145	..	..	..	4 32	..	1041
Newton-Stewart lev.	..	..	8 58	..	1152	1150	..	..	..	4 40	..	1051
Creetown	..	..	9 11	..	12 7	12 3	..	..	..	4 53	..	..
Gatehouse-of-Fleet	..	..	..	..	1219	1213	§	..	..	..	..	..
Lochskerrow	..	..	9 35	..	1231	1224	..	..	..	5 15	..	..
New Galloway	..	..	9 44	..	1240	1233	..	..	..	5 25	..	..
Parton	..	..	9 50	..	1246	1239	..	..	..	5 33	..	..
Crossmichael	..	..	9 57	..	1253	1246	..	..	..	5 40	..	..
Castle-Douglas arr.	..	..	10 3	..	1259	1252	..	..	..	5 46	..	1146
Kirkcudbright lev.	6 45	8 30	9 35	..	1222	..	..	1 53	4 51	..	..	..
Tarff	..	8 38	9 43	..	1230	..	..	2 1	5 0	..	..	..
Castle-Douglas arr.	7 2	8 49	9 54	..	1241	..	..	2 12	5 11	..	..	..
Castle-Douglas lev.	7 3	8 51	10 5	..	1 2	1255	..	2 15	5 12	5 52	..	1152
Dalbeattie	7 13	9 0	1016	..	1 15	1 8	..	2 25	5 21	6 6	..	..
Southwick	..	9 6	..	..	1 21	..	..	2 31	5 27	..	..	..
Killywhan	7 25	9 14	..	..	1 29	..	..	2 41	..	..	..	..
Dumfries arr.	7 38	9 27	1041	..	1 43	1 33	..	2 56	5 48	6 30	..	1229a
13 CARLISLE	..	1035	1158	..	3 4	3 30	..	5 48	6 35	8 15	..	1a20
13 GLASGOW (St. Enoch) arr.	10 2	..	2 28	..	3 53	3 53	..	5 32	7 45	9 0	..	7N30k
62 EDINBURGH (Wav.) K	1136	..	4 30	..	5 52	5 52	..	7 11	9 30	1043	..	9H0

(Last column: The Northern Irishman TC and SC Stranraer Harbour to London (Euston) arrive 8-10 a.m.)

* Saturdays only
§ Mondays only
† Except Saturdays
‡ Except Mondays
A Passengers leave Portrush 3-50 p.m.
A Mondays and Saturdays 2 minutes later

k Leave Dumfries for Glasgow at 4-8 a.m.; Sats. arrive 7-21 a.m.
K Via Stranraer; leave 5-15 p.m. on Saturdays. Passengers find own conveyance from Town to Harbour Station
K Via Glasgow. Change between Queen Street and St. Enoch Stations.
L Leave Stranraer Harbour 6-30 a.m.
M Passengers arrive Portrush 12†43 p.m. 12*56 p.m
N On Sunday mornings leave Dumfries 6-25 a.m., arrive Glasgow (Central) 8-30 a.m. (until 22nd Sept., 1957, inclusive) and from 18th May, 1958, arrive Glasgow (St. Enoch) Mondays and Saturdays only
SC Sleeping Cars ‡ Fridays only TC Through Carriages

The Murray's Railway Guide passenger timetable for April 1958.

boat train service now ran from Euston instead of St Pancras (though strangely, in the up direction, by changing at Carlisle one could arrive at St Pancras earlier than at Euston).

After the transfer to LMS management in 1923 there were no changes to the service pattern for some while, except that the boat trains were somewhat accelerated. By 1931 the arrival time at Belfast had been brought back to 9.10 am; four years later it was 8.58. In 1931, too, third class sleeping cars were included in the boat train formations. A further innovation during that summer was a through portion from and to Newcastle-upon-Tyne, attached or detached at Carlisle. Going westward it left Newcastle in the small hours, and in the opposite direction arrived similarly. This facility continued until the outbreak of World War II.

Again, war-time exigencies brought about a reduction of services, and by the end of 1947 these had not much recovered, one boat train and two stopping trains running in each direction. The first and third class sleeping cars, however, were again available. After the change to BR management the through night service to Newcastle came back, though only at week ends on an up Friday night and down Saturday morning service. By 1950 there were again four trains each way along the whole line, including a mid-day service from and to Carlisle connecting with a day boat to and from Larne.

By 1952, however, the service had somewhat deteriorated. The local 'wee train' had ceased to run, as also had the through carriages to and from Newcastle. In 1956, however, the latter came back, while the daytime boat trains originated also at Newcastle instead of at Carlisle. Then, seven years later, as already recounted, the axe fell, and the railway ceased to exist, partly through official decision but partly also because of discontinuance of custom.

It remains to give an account of the fortunes of the Whithorn branch – the former Wigtownshire Railway. In 1877, when it was finally extended to Whithorn, four trains a day ran in either direction, which took round about an hour for the journey, making a connection at Millisle from or to Garlieston. In 1880 an extra train was included, of coaches attached to the goods train which ran each way daily. This took an unconscionable 3 hours to cover the 20 miles from one end of the line to the other and cannot have enjoyed much patronage. It was still running in 1885 but had vanished from the timetable five years later. In 1891 the number of daily ordinary passenger trains had increased to six southbound and five northbound, all running during the morning and evening with a long gap between. In 1893 the extra southbound train was taken off and the five-each-way pattern persisted till 1900, when one of the northbound trains was dropped, though there was an extra train each way on Friday. The pattern remained thus until the outbreak of World War I. Over the years the schedules quickened slightly; by 1914 some trains were doing the whole journey in 50 minutes and one even managed it in 44 minutes.

The War had its effect; by 1918 the service had come down to four southbound and three northbound trains; by 1923 an extra journey had been added in each direction, with extra Friday trains being run to fill the afternoon gap; this pattern now persisted until the eve of World War II. It is curious that while a connection from Newton Stewart was afforded to passengers from the early morning train from London, none was provided for those going in the other direction.

Most unusual of all the services provided was a summer operation begun in 1928, of through carriages between Whithorn and Edinburgh. Departure from Whithorn was at 9.25 am; at Newton Stewart the vehicles were attached to the morning stopping train from Stranraer and detached at Dumfries. They were then taken through over the former CR line to Lockerbie and there coupled to an express from the south, which reached Edinburgh at 3.15 pm. In the opposite direction a similar portion left Edinburgh at 11.45 am (in some years at 10.05 am) and arrived at Whithorn at 5.41 pm. The service was still running in 1937, but ceased soon afterwards. There cannot have been many passengers who used it the whole way.

The coming of World War II reduced the Whithorn branch's service to four southbound and three northbound trains daily, and it remained at this level until the line was closed to passenger traffic in the autumn of 1950.

Two 4–4–0s No. 14340 (Dunalastair III) and No. 14192 (Pickersgill) at the head of the 12.30 pm service to Glasgow on 11th August, 1930. The coach behind the locomotives appears to be a former LNWR dining car. *H.C. Casserley*

Chapter Fifteen

Accidents and Mishaps

The Portpatrick Railway had its share of accidents, a few of which were serious, during the hundred years of its existence, but only four caused human fatalities, and on no occasion was a passenger killed, which was not a bad record for 65 miles of line in existence for a whole century. On the branch to Whithorn, indeed, no accident or serious mishap seems to have occurred at all.

The first unfortunate incident was the death of the station master at Dromore (later Gatehouse of Fleet) on New Year's Day 1863. A married man with a family, and newly appointed to this lonely post, he had taken the evening train to Creetown and spent a convivial evening with friends, perhaps indulging a little too freely (though the local paper noted that when he left to return home he was still able to walk. It would have been better for him if he had *not* been able to do so). He chose to go back along the line, a little under five miles, but never reached his home. The next day he was found badly injured beside the line a mile east of Creetown station, having apparently been struck by the evening train from Stranraer, and died from his injuries soon afterwards.

On 8th December, 1866 there was an accident to the morning down goods train sufficiently serious to warrant a Board of Trade investigation; this too happened near Creetown. The train, headed by CR No. 265 (formerly the Portpatrick Railway's No. 4, a mixed-traffic 0–4–2) with 17 wagons and a brake van, was descending the 1 in 80 gradient north-west of the station without any brake applications being made. Reaching a curve ½ mile short of Palnure, the engine left the rails and the whole train plunged down an embankment. The tender parted company with the locomotive and turned round and half over on to its left side, crushing the unlucky fireman to death. The train vehicles distributed themselves in a heap beside the locomotive; the driver extricated himself with a few bruises while one of the two guards had a fractured arm. At the Inquiry the cause was found to be excessive speed, but the condition of the track also came in for criticism, since it had been constructed without fishplates to fix the rail ends firmly together.

In the small hours of Sunday morning, 6th October, 1872, a similar accident befell a westbound goods train between Glenluce and Dunragit, which served to highlight the poor conditions under which the company's employees worked. At the foot of the 1 in 83 gradient beyond Glenluce the train became derailed, and the brake van was destroyed in the subsequent pile-up, the guard being killed. Records showed that he had been continuously on duty since the previous Friday evening! At the time of the accident he had clearly been awake and alert, despite his fatigue, since he had noted the time of passing Glenluce in his log-book. Only three days earlier he had been one of a group of guards who had signed a petition for an increase of wages.

On 16th June, 1873 a more serious accident occurred, though happily without fatalities. The afternoon passenger train from Castle Douglas to Stranraer had been halted at Creetown to cross with an eastbound train. It had been standing there for 10 minutes when a goods train, which had left Castle Douglas ¾ hr before it, and which had been overtaken by the passenger train at the Loch Skerrow loop, suddenly appeared round the curve, having ignored adverse signals, and ran into the back of it, injuring 11 people. It appeared from evidence given at the subsequent Inquiry that the driver, fireman and first guard, who were all travelling on the footplate, had spent part of their time while shunting operations were being performed at Crossmichael in drinking at the local hostelry. After that their behaviour became increasingly reckless. The driver (whom the accident appeared to have sobered up) simply absconded; the fireman and guard were dismissed.

Later the same year, on 3rd November, an equally serious accident occurred at Newton Stewart, though the fault this time lay with the inadequate maintenance of the signalling system. An up goods train from Stranraer reached the station at 7.25 pm and waited there while its locomotive was involved in some shunting. A down passenger train was due, and the pointsman had duly put both the down home and distant signals 'on' to cover the goods locomotive's operations. Unfortunately the wire to the distant signal had been allowed to become slack, and in addition one of its levers was sticking, so that it showed the customary white light, the equivalent of the later green. The driver of the passenger train naturally assumed from the distant signal that the line was clear, and his view of the home signal (which *did* show red) was obscured by steam both from his own locomotive and that of the goods train, so that he struck the other engine as it was blocking his path. Thirteen passengers and the guard received injuries, some of them severe.

A 20 years' intermission from accident followed – apart from a minor incident in September 1879 when a shunting engine bumped into the end of a passenger train on Stranraer pier, causing damage to two buffers and slight injuries to two passengers – until the winter of 1894–1895, when there was a severe storm during December and a blizzard in February. The December storm, though it played havoc with communications between Stranraer and Larne, and damaged one of the moored packet-vessels, *Princess May*, did no more than delay the main rail services and actually stopped one train in its tracks. The morning train from Stranraer to Portpatrick on Saturday 22nd December, while approaching Piltanton viaduct two miles south-west of Stranraer, experienced so much side pressure from the wind that it stopped short of the arches, and the footplate crew, guard and passengers held a council of war on the spot. All agreed it would be foolish to attempt to go further and the train accordingly backed down the gradient to Stranraer. The same morning the boat train, running in two portions, only just managed to get through, braving the hazard-spots of the Wee Fleet and Big Fleet viaducts; the second portion was much later in arriving at Stranraer than the first, since it had been delayed at New Galloway station for an hour while the signal box, which had been blown on to the line, was cleared out of the way.

The snowstorm which blew up on 6th–7th February blocked the cuttings in several places along the line. No trains ran on the Whithorn branch for five days, and none on the Stranraer–Portpatrick branch for six. On the main line from Castle Douglas to Stranraer the afternoon train on the 6th managed to complete its journey; others in both directions became snowbound. The 7.28 pm from Dumfries got as far as Loch Skerrow and was then marooned there for several days, the train crews and two passengers (all there were) managing to reach the nearby platelayers' cottages, where their appetites put such a strain on their hosts' larders that it became necessary to return to the train and raid one of the vans which was carrying supplies of food as part of its cargo.

In the opposite direction the 3.40 pm from Stranraer became stuck on the 1 in 80 gradient a little to the north of Creetown, where after making efforts to free itself the engine eventually ran out of steam. An attempt to clear the line from the east with a snowplough failed. A whole night and day passed before the storm abated sufficiently to allow passengers to be taken from the train to the station buildings at Creetown, where the accommodation was at least less cramped and warmer and some hot food could be had. After a while most managed to make their ways down the hill to the village, where they were able to find accommodation in local inns and guesthouses. Not until Saturday afternoon were they able to entrain again and resume their journey. One passenger was so upset by the whole experience that he attempted to cut his throat, but he was discovered in the act, the razor was snatched away from him and medical attention was obtained. The line was not fully opened for traffic until Monday 11th February.

Another decade followed free from untoward incidents, until an unfortunate accident, on 28th December, 1905, at Loch Skerrow, when the signalman, as he was actually putting the tablet in place on its post, was struck on the face by the extended tablet-catcher on the locomotive of the up evening express from Stranraer to Carlisle. This was the last occasion when anyone was killed on the line, though strictly speaking not the last fatality, for on 16th March, 1907 the line between Gatehouse of Fleet station and the Big Fleet viaduct was blocked by a fall of rock. An up goods train carrying cattle and sheep ran into the obstruction and was derailed. Sixty of the sheep were so badly injured that they had to be slaughtered on the spot. Twelve years later a number of cattle were also killed on the line, as will be mentioned later.

The War years saw no accidents, though accidents were in the making through the great use made of the line and the lack of necessary maintenance. The first sign of how seriously the state of the track had deteriorated appeared when an empty van on the evening westbound passenger train from Dumfries came off the rails a little east of Glenluce. The consequence was the breakage of 2,330 chairs and 159 sleepers before the train was finally brought to rest. This was on 30th January, 1918; 3½ years later, almost at the same spot, on 1st August, 1921, on the same train, an empty meat van derailed itself, was dragged for some distance, finally broke free and straddled across the line. Official concern caused an investigation to be ordered, but it does not appear that one was carried out; if it was the findings were not published.

Fifteen months later, right at the end of the Joint Committee's separate existence, a much more serious accident occurred when the 3.40 pm passenger train from Stranraer to Dumfries was derailed as it crossed the Cree viaduct at a speed of about 60 mph, just before stopping at Palnure. A four-wheeled van at the front of the train left the rails, and it and the four following vehicles were overturned and wrecked. None carried any passengers, who were in the coaches which followed and which were not derailed, so fortunately there was no loss of life. A curious thing about the accident was the fact that the driver and fireman were both unaware of what had happened until a passing member of a permanent way gang called up to them that they had no train behind them. The van's coupling had snapped, but in the darkness (it happened in late December) the engine crew had not noticed that they were going on alone. Major Hall, who carried out the Enquiry for the Ministry of Transport, put the blame on excessive speed and recommended a limit of 45 mph should be imposed over the bridge itself. What was really needed, and under LMS management eventually came, was a thorough renewal of the track, which the passage of innumerable trains during the War years and the lack of proper maintenance had rendered dangerous.

Only one serious accident occurred after the 1923 grouping when the line became part of the LMS system. On 30th December, 1935 the early morning mail and boat train connecting with the Stranraer–Larne Steamer, made up of seven coaches, which included Post Office vehicles and first and third class sleeping cars, hauled by two class '2' 4–4–0s, became derailed as it was leaving the Ken viaduct. All the coaches rolled down the embankment; the two leading ones, carrying the mails, were completely destroyed, and bewildered Post Office workers found themselves sprawling in the grass among the debris. Astonishingly none of them was hurt, and only two passengers suffered minor injuries, probably because of the sturdy construction of the five passenger vehicles. The cause of the derailment was never definitely established, but suspicion fell on the pilot engine, whose tender was believed to have left the rails first. A second derailment suffered by the same locomotive shortly afterwards, when it was returning from Portpatrick to Stranraer, going tender first, caused it to be regarded as an unlucky engine, and it was shortly afterwards sent south of the Border to a part of the LMS system where its proneness to accident was not known.

One should also mention two serious incidents which occurred within days of each other. The 24th March, 1919 was a day of high winds which persisted into the evening. Two locomotives were stationed for the night inside the shed at Newton Stewart; they normally worked on the Whithorn branch during the daytime. A cleaner put a lighted lamp down on an oil-soaked bench; its flickering flame set the bench alight, and soon the whole shed was burning. The fire reached the coal in the engines' tenders, which were soon red hot as the wind from outside reached them when the wooden walls of the shed had burned down. After their ordeal the engines, one of Smellie's 'Wee Bogies' and a rebuilt 0–4–2, needed very extensive repair. A few days later, 27th March, a herd of valuable Highland cattle strayed on to the line near Creetown through an open gate and wandered down the track munching the grass at the lineside. The morning boat train, behind a Manson 4–4–0, was stopped at Creetown and the driver was told that there were cattle on the line. For two miles the train crawled along, but no sign of them was seen and it was supposed that they must have left the line again. So speed was increased. Then, almost at the foot of the incline, the locomotive suddenly ploughed right into them. Thirteen of them were killed and one of the coaches was derailed.

All these accidents and mishaps fade into insignificance when contrasted with the sinking of the packet vessel *Princess Victoria* when it sank with all hands and most of its passengers during a storm on 31st January, 1953. This, however, did not happen on the railway and so is described separately in Appendix 6 (*page 142*).

The scene of the 1935 accident at Parton, near Castle Douglas, to the Stranraer boat train, in which no serious casualties occurred. *Courtesy Railway Gazette*

A view of Castle Douglas station in 1956. *Oakwood Collection*

A scene at Wigtown with the last train (freight only) having just arrived in 1964.
Stranraer Public Library

The last train, headed by No. 55125, 0–4–4 ex CR tank leaving Portpatrick in 1950.
Stranraer Public Library

Chapter Sixteen

The Disused Line and the Future

The fate of the track-bed of the Portpatrick and Wigtownshire lines, once the trains ceased to run and the rails had been lifted, was as with most other closed lines bound to be a reversion to nature unless some decision were made to use it for some other purpose, such as turning it into a walk-way or cycle-way (as has now happened with the first 2½ miles of the former Callander and Oban Railway) or converting it into a road (as has been the case with part of the former Caledonian Railway's approach to Princes Street Station in Edinburgh). No such modification has so far occurred in Galloway, which in view of recent developments (mentioned below) may be just as well.

Some of the buildings associated with the line have been converted to dwelling-houses, as at New Galloway station. Others have for the most part disappeared – particularly if they were made of wood, or of steel which could be sold for scrap. Newton Stewart station has vanished as if it had never been there; the site is now occupied by a shell-fish processing factory and other such enterprises. A former shed still stands, in the angle between where the main line and Whithorn branch forked, and appears to be a store for hay. Something still remains of Stranraer Town station, including the frontage on to the road. As to the bridges, a few are still standing, and one (the Glenluce viaduct) can be traversed. Piltanton viaduct was blown up when the rails between Stranraer and Colfin were lifted and only a few overgrown fragments of masonry remain. The Cree bridge was dismantled, though parts of the timber abutments may still be seen, and more is visible at low tide. The Wee Fleet viaduct was recently destroyed in an Army exercise, and the Big Fleet viaduct will probably soon be similarly demolished. The Ken viaduct seems solid enough still; situated where it is, it would seem a good idea to build a single-track road across it and charge users a small toll, since there was once a ferry at this point.

Attempting to walk the whole length of the line is not an enterprise to be commended at the present day. In 1974, less than 10 years after closure, it was possible, and in 1975 three successive issues of the *Scots Magazine* published an account by K.M. Andrew who walked the stretch from Dumfries to where Challoch Junction had been – not, however, including the Whithorn or Portpatrick branches. He found that much of the trackbed was overgrown, and was occasionally informed that he was trespassing. Nearly all the bridges were still usable, but the Cree bridge had gone. At the time of writing this book (1990) that was 16 years ago, and probably the only part that can still be walked with enjoyment is the section from New Galloway station to Gatehouse of Fleet station, which for most of its route passes through forestry plantations; however, it would be necessary to wade through both the Big and Wee Waters of Fleet.

The writer personally attempted to walk along three short sections. One was from Newton Stewart to the site of the Cree bridge, which is still, with occasional diversions, generally traversable. Another was the first four miles of the Whithorn branch, from Newton Stewart to Causewayend; this presented several difficulties through bramble-filled cuttings. A third was the Portpatrick branch, which I tried to follow from the point where it crossed the Stranraer–Drummore road as far as Colfin. It proved impossible, on a wet day, to cross the Piltanton burn, and being short of time I gave up the attempt to go further. A motorist who gave me a lift back to Stranraer told me that from Portpatrick to Colfin it is negotiable and that the Local Authority was considering making a walkway of it, which one hopes is correct. Generally, it seems to me, in regard to the whole line, that if one wishes to explore it it is better to use a car or bicycle and approach it from a number of different points. Much of it has now been purchased by local farmers and it is questionable whether any rights of way exist.

A much more hopeful prospect than the conversion of the old line to roads or walkways is at the time of writing opening up. Could the line be re-opened? Re-opening is certainly being envisaged. There is no doubt that better communications are needed between Northern Ireland and Great Britain, particularly on this side of the North Channel. At present, if one travels by rail, one has either to go by the circuitous route through Kilmarnock and Ayr when going to or from the South, or one has to take the A75 road which excessive lorry traffic is rendering unpleasant. (Some 15 ships sail daily between Loch Ryan and Larne in either direction conveying all kinds of motor vehicles of all sizes.) Local authorities both in Scotland and Northern Ireland are pressing for the reintroduction of a direct rail link, and a meeting was recently held at Stranraer between local authority representatives and railway representatives, both British and Irish, which unanimously urged the funding by the Government of a feasibility study, whose cost was estimated at perhaps £50,000 – mere chicken-feed when so much is at stake. As to the complete cost of the project, a figure of £500 million was put forward as likely – not, I consider, excessive in these days when projects are measured in billions rather than millions.

Were the project to mature, one might well expect a different route to be followed from the one built 130 years ago. It would seem sensible now, if the cost were not considered prohibitive, to take in the two burghs which were previously passed by; the line of the former trackbed from Castle Douglas to Kirkcudbright could be used as far as the village of Tongland, with the 'Parkway' type of station being established there to serve the near-by burgh. The route would then not need to cross the Dee estuary but could cut straight across the Tarff burn towards Gatehouse, through a station placed just north of that town, and then continue up the valley of the Water of Fleet to re-join the former route at the site of the former Gatehouse station. The additional cost of building these 14 miles across country where no railway was constructed previously could be set off against the fact that the Ken, Wee Fleet and Big Fleet viaducts would not need re-building. Some bridges would have to be constructed or reconstructed, of course, especially that across the Cree, and some re-routing would probably be necessary near Newton Stewart, where the line could parallel the present by-pass road to the south of that town, with a station being built a mile or so away from it. The Glenluce viaduct still looks strong enough to take a single line of railway.

A few years ago such an undertaking would have seemed a mere pipe-dream, but if it could be achieved it ought now to be possible to run the line more economically than was feasible in 1965. The long detour between Dumfries and Stranraer by way of Kilmarnock and Ayr would no longer be necessary. Radio-signalling, now a proved success on other Scottish single-track lines, would be far less expensive than the old method. Stations need only be unstaffed halts with car parking facilities. Sprinter units would be cheaper and quicker to operate than locomotive-hauled trains for passenger traffic. If tunnels could be avoided and the loading gauge were made high and wide enough, outsize containers, or perhaps even whole lorries taken on flat trucks on a 'piggy-back' system, could be transported, so that less money would be needed to improve the overcrowded A75 road. Money from the European Economic Community would probably be available to finance the enterprise. One is bound to wish the project well, for more than sentimental reasons.

Appendix One

Principal Bridges on the Line

Situation	*Description*	*Present Condition*	*Date*	*Contractor*	*Grid Ref.*
Loch Ken Viaduct: West of Parton, across narrowest part of L. Ken. For full description see note below.	3 bowstring girder spans with masonry piers and abutments.	Still standing and can be crossed on foot, with care.	1860	Thos Nelson & Co.	2684 5703
Loch Stroan Viaduct: 1 mile west of New Galloway station, across Black Water of Dee.	4 arches, piers and abutments, of masonry.	Still standing and can be crossed on foot, with care.	1860	McNaughton & Waddell	2645 5701
Little Water of Fleet Viaduct: 2 miles west of Loch Skerrow passing place.	9 masonry arches of local granite.	Now demolished.	1860	McNaughton & Waddell	2587 5671

Big Water of Fleet Viaduct: 5½ miles west of Loch Skerrow passing place.	20 masonry arches of local granite, piers reinforced later by brickwork sheathing.	Still standing but likely to be demolished shortly.	1860	McNaughton & Waddell	2560 5644
Culcronchie Bridge: Across Culcronchie Burn, 3 miles west of Gatehouse of Fleet station.	4 masonry arches.	Now demolished.	1860	James Gowans & Co.	2506 5613
Graddoch Viaduct: Across Graddoch Burn, 2½ miles north west of Cree-town station.	8 masonry sand-stone arches.	Partly demolished.	1860	James Gowans & Co.	2462 5631
Palnure Bridge: Across Palnure Burn, ¼ mile short of Palnure station.	3 girder spans on sandstone piers, across Palnure Burn.	Still standing.	1860	James Gowans & Co.	2455 5634
Cree Bridge:	3 girder spans on timber piers with stone abut-ments across tidal reach near mouth of R. Cree.	Demolished.	1860	McDonald & Grieve	2436 5635
Upper Bladnoch Bridge: Across R. Bladnoch 4½ miles west of Newton Stewart station.	Single span bowstring girder bridge with stone abutments.	Demolished.	1860	McDonald & Grieve	2342 5629
Glenluce Viaduct: Across Water of Luce ½ miles west of Glenluce station.	8 arches on brick piers.	Still standing and can be crossed on foot with care.	1860	James Falshaw & Co.	2192 5573
Piltanton Viaduct: 2½ miles south of Stranraer, across the Piltanton Burn.	13 brick arches on masonry piers.	Demolished.	1861	James Falshaw & Co.	2059 5575
Lower Bladnoch Bridge: 1 mile south west of Wig-town station, across R. Bladnoch 1½ miles from its mouth.	2 girder spans on masonry abutments and central pier.	Demolished.	1875	John Granger	2425 5541

The bowstring arches of the massive Ken viaduct across the 'narrows' of Loch Ken.
Julia Cheetham

The Ken Viaduct: A Contemporary Description

The *Dumfries Courier* of 1st September, 1859 included a description of this bridge, written while it was in process of construction, from which the following has been extracted:

> The main structure will consist of three arches, each of a span of 130 feet clear, besides two small arches on each side of the bank . . . The piers are elegant in form, and are substantially built of reddish sandstone from Cove and Cumberland, the stones being blocked out before being despatched to the Dee. There will be two piers in the deep waters of the river, and owing to the great depth of water these piers will be partially built within cylinders of iron. These cylinders have been brought to the spot, and those intended for one of the piers placed in the river. The cylinders are each six feet in height and about the same in diameter. Six of them placed one above the other reach nearly to the top of the water . . . the lower sinking down from the superincumbent weight into the sand and gravel of the bed of the stream. The cylinders are firmly joined together and form an elongated coffer dam which will be permanent . . . A pump driven by steam power is placed on the staging above and ready to commence clearing the cylinders of water. As soon as this has been effected a foundation of concrete will be laid and building within the cylinders commenced. It has already been mentioned that there will be two piers in the water; but each of these piers will be double below the water, and for each two sets of cylinders will be required. When the piers within the cylinders have been filled with masonry they will be joined by an arch, and a single large pier resting on the arch will be raised above the waterline. The cylinders will remain around the lower piers and under the water, so that the piers will appear to be wholly of stone. Rising from the summits of the piers, which will be 50 feet high, semicircular iron girders will stretch their huge arms high above the water, to which they will be concave, and from them will be suspended the iron tramway for the railroad . . . The diameter of the main piers will be 8 feet . . . This promises to be not only a stupendous but a most elegant erection and must prove an ornament to the beautiful scenery about the Boat of Rhone . . .

This description may be compared with the appearance of the bridge as shown on the previous page.

A distant view of the magnificent Big Water of Fleet viaduct. *Julia Cheetham*

Appendix Two

Mileages

1. Main Line

	Miles
Castle Douglas	0.0
Crossmichael	3.6
Parton	6.8
New Galloway	8.9
Loch Skerrow loop	12.9
Gatehouse of Fleet	18.9
Creetown	23.5
Palnure	26.6
Newton Stewart	29.9
Kirkcowan	36.1
Glenluce	43.5
Challoch Junction	45.9
Dunragit	46.9
Castle Kennedy	49.7
Stranraer Harbour Junction	51.9
Stranraer Town	53.1

2. Portpatrick Branch

	Miles
Stranraer Town	0.0
Colfin	4.0
Portpatrick	7.4

3. Whithorn Branch

	Miles
Newton Stewart	0.0
Causewayend siding	3.6
Wigtown	6.9
Kirkinner	9.3
Whauphill	10.9
Sorbie	13.2
Millisle Junction	15.2
Whithorn	19.1

4. Garlieston Branch

	Miles
Millisle Junction	0.0
Garlieston	1.0
Garlieston Pier	1.2

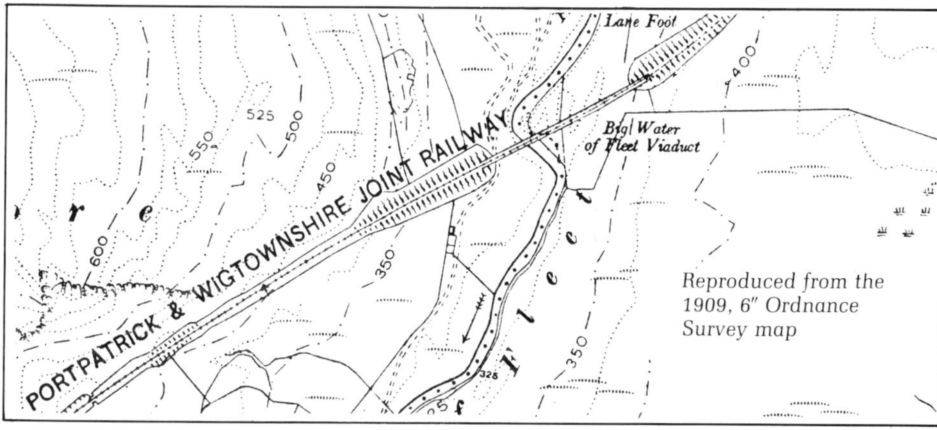

Reproduced from the 1909, 6" Ordnance Survey map

Close up detail of the massive brick pillars and strapped masonry of the Big Water of Fleet Viaduct. *Julia Cheetham*

Appendix Three

Principal Events on the Portpatrick and Wigtownshire Railways

	Date
Act of Parliament establishing the Portpatrick Railway	17/8/1856
Opening of Portpatrick Railway: Castle Douglas to Stranraer	12/3/1861
Opening of continuation: Stranraer to Portpatrick	28/8/1862
Caledonian Railway assumed working of line	1/10/1864
Act of Parliament establishing the Wigtownshire Railway	18/7/1872
First carriage of mails to and from Ulster by Portpatrick Railway	1/2/1875
Opening of Wigtownshire Railway to Wigtown	7/4/1875
Thomas Wheatley appointed to manage Wigtownshire Railway	31/7/1875
Continuation of Wigtownshire Railway to Garlieston completed	2/8/1875
Completion of Wigtownshire Railway to terminus at Whithorn	7/7/1877
Death of Thomas Wheatley	13/3/1883
Junction with line from Girvan made at Challoch	1/8/1883
Formation of Portpatrick and Wigtownshire Joint Committee	6/8/1885
First through carriage service from London to Stranraer	1/11/1885
Inauguration of summer 'daylight crossings' between Stranraer and Larne, with connecting trains	14/7/1890
The 'Great Snowstorm' in Galloway	6/2/1895
Closure of Garlieston branch to passenger traffic	1/3/1903
Introduction of steam road motors from Stranraer to Drummore	1/5/1907
Outbreak of World War I	4/8/1914
End of World War I	11/11/1918
Transfer of Portpatrick and Wigtownshire Joint Committee to LMS Railway	1/1/1923
Introduction of 1st Class sleeping cars: Euston to Stranraer	1922
Introduction of 3rd Class sleeping cars: Euston to Stranraer	1931
Outbreak of World War II	3/9/1939
Completion of Cairnryan Railway	7/1943
End of World War II (in Europe)	7/5/1945
Formation of British Railways	1/1/1948
Closure of Portpatrick Branch to passengers	6/2/1950
Closure of Wigtownshire Branch to passengers	23/9/1950
Closure of Stranraer–Colfin section to all traffic	1/4/1959
Closure of Wigtownshire Branch to all traffic	5/10/1964
Final closure of former Portpatrick Railway	12/6/1965

An 0–6–0T, No. 16351, near Stranraer in April 1948 still in LMS livery, used for shunting duties in the area. *H.C. Casserley*

The fourth *Princess Victoria* seen here off Stranraer in 1953 shortly before it foundered in a storm, with loss of all hands and many unfortunate passengers (*see page 142*). *Stranraer Public Library*

Appendix Four

Principal Events on the Cross-channel Steamship Services to Ulster

	Date
Stranraer–Belfast service begun by SS *Briton*	4/12/1865
. . . and ended	30/1/1868
Short Sea Steam Packet Company began Larne–Stranraer service with SS *Fannie* and SS *Alice*	13/7/1868
. . . and ended	31/10/1868
Larne and Stranraer Steamboat Company began service with *Princess Louise*	1/7/1872
First voyage of *Princess Beatrice*	3/2/1876
First voyage of *Princess Victoria I*	24/4/1890
Disposal of *Princess Louise*	6/1890
First voyage of *Princess May*	9/5/1892
First voyage of *Princess Maud*	1/6/1904
Disposal of *Princess Beatrice*	1904
Disposal of *Princess Victoria I*	23/2/1910
First voyage of *Princess Victoria II*	29/4/1912
Requisitioning of *Princess Victoria II* for military uses	10/1914
Disposal of *Princess May*	10/12/1914
First voyage of *Princess Margaret*	6/1931
Disposal of *Princess Maud*	1931
Disposal of *Princess Victoria II*	3/1934
First voyage of *Princess Maud II*	3/1934
First voyage of *Princess Victoria III*	7/1939
Requisitioning of *Princess Victoria III* and *Princess Maud II* for military uses	9/1939
Sinking of *Princess Victoria III* on war service	5/1940
Temporary transference of *Hampton Ferry*, *Shepperton Ferry* and *Twickenham Ferry* to Stranraer for carriage of vehicles	7/1940
First voyage of *Princess Victoria IV*	3/1947
Foundering of *Princess Victoria IV* between Stranraer and Larne	31/1/1953
First voyage of *Caledonian Princess*	12/1961
Disposal of *Princess Margaret*	3/1962

Appendix Five

Dimensions and Details of CR, G&SWR, LM&SR and BR locomotives used on the Portpatrick and Wigtownshire Railways' routes: 1864–1965

Col.	**1**	**2**	**3**	**4**	**5**	**6**	**7**	**8**	**9**	**10**	**11**	**12**	**13**	**14**
Type	CR 2–4–0	G&SW 2–4–0	G&SW 4–4–0	CR 0–6–0	G&SW 4–4–0	CR 4–4–0	G&SW 4–4–0	CR 4–4–0	4–4–0 Compound	LMS 4–4–0	LMS 2–6–0	LMS 4–6–0	LMS 4–6–0	BR 4–6–2
Designer	B. Connor	H. Smellie	H. Smellie	D. Drummond	J. Manson	D. Drummond	P. Drummond	W. Pickersgill	R.M. Deeley (modified by H. Fowler)	H. Fowler	G. Hughes	W. Stanier	W. Stanier	R.A. Riddles
Yr of Bdg	1869/71	1879	1882	1883/97	1904/6	1884/94[b]	1915	1916/21	1924/27	1928	1926/32	1934/51	1934/36	1951/52
Total Heating Surface (sq.ft)	982.2	1206	1065	1202.3	1434	1404	1923[c]	1529[c]	1607.7	1410.2[c]	1812[c]	1846[c]	1853[c]	1878[c]
Grate Area (sq.ft)	14.3	16.0	16.0	19.5	18.2	20.5	27.5	20.7	28.4	25.0		27.8	29.5	36
Boiler pressure (lb./sq.in)	140	140	140	150	170	160	180	180	200	180	180	225	225	225
Cylinders (in.)	17 × 22 (2)	18 × 26 (2)	18¼ × 26 (2)	18 × 26 (2)	18¼ × 26 (2)	18 × 26 (2)	19½ × 26 (2)	20½ × 26 (2)	HP 1 19×26 LP 2 21×26	19 × 26 (2)	21 × 26 (2)	18½ × 28 (2)	17 × 26 (3)	19½ × 28 (2)
Coupled Wheel dia.	6ft 2in.	6ft 9½in.	6ft 1in.	5ft 0in.	6ft 9½in.	6ft 6in.	6ft 0in.	8ft 6in.	6ft 9in.	6ft 9in.	5ft 6in.	6ft 0in.	6ft 9in.	6ft 2in.
Coupled wheelbase	8ft 7in.	8ft 3in.		16ft 3in.	8ft 9in.	8ft 0in.	10ft 0in.	9ft 9in.	9ft 6in.	9ft 6in.	16ft 6in.	15ft 0in.	15ft 4in.	14ft 0in.
Total wheelbase	15ft 2in.	16ft 3in.		16ft 3in.	21ft 8in.	21ft 1in.	24ft 9in.	24ft 4in.	24ft 3in.	22ft 8½in.	25ft 6in.	27ft 2in.	27ft 7in.	35ft 9in.

Appendix Five

Dimensions and Details of CR, G&SWR, LM&SR and BR locomotives used on the Portpatrick and Wigtownshire Railways' routes: 1864–1965

Col.	1	2	3	4	5	6	7	8	9	10	11	12	13	14
Wt of locomotive in wkg order	34.98 tons	38.5 tons	41.5 tons	41.25 tons	50.4 tons	46.05 tons	61.85 tons	61.25 tons	61.7 tons	54.05 tons	66.0 tons	74.0 tons	79.55 tons	88.5 tons
Adhesion weight		26.9 tons	27.75 tons	41.25 tons	30.85 tons	30.4 tons	42.0 tons[d]	39.75 tons	39.2 tons	34.95 tons	56.15 tons	54.0 tons	60.0 tons	56.9 tons
Tractive effort[a] lb.	10,224	12,376	14,116	17,901	15,353	14,688	21,009	21,432	22,649	17,729	26,580	25,455	26,610	27,516
Details of Tender:														
Coal cap.	4 tons	3¾ tons	4 tons	4½ tons	4 tons	4 tons	6½ tons	6 tons	5½ tons	4 tons	5 tons	9 tons	9 tons	7 tons
Water cap.	1500 gal.	2100 gal.	2200 gal.	2840 gal.	2500 gal.	2840 gal.	3800 gal.	4200 gal.	3500 gal.	3500 gal.	3500 gal.	4000 gal.	4000 gal.	4250 gal.
Weight in wkg order	19.26 tons	27.0 tons	27.0 tons	34.0 tons	32.25 tons	34.85 tons	47.3 tons	46.5 tons	42.7 tons	41.2 tons	42.2 tons	52.0 tons	54.65 tons	49.15 tons

Notes: (a) At 85% of boiler pressure.
(b) As subsequently rebuilt by J. McIntosh.
(c) Including superheater.
(d) Estimated: no figure was published.

Appendix Six

The Loss of the Princess Victoria in 1953

On 31st January, 1953 the *Princess Victoria* left Stranraer at 7.45 am on her last crossing, one which she never completed. It was the stormiest day around the coasts of north-west Europe that anyone could remember. However, the weather forecast prophesied a moderation of the wind, and Captain James Ferguson seems not to have felt that he should remain in port; his ship had been only recently built and was considered seaworthy. The passage would be bumpy and no doubt uncomfortable, but this would be nothing new; the return voyage would no doubt be a little calmer. No cars were carried as the force of the wind prevented them from being loaded, but a certain amount of cargo was loaded and placed on the empty car deck.

As far as the exit from Loch Ryan there were no problems. The ship went in and out of squalls of sleet and snow before being lost to sight from the land, and took a course some miles to the north of her usual track before turning towards Larne to allow for the wind driving her south. From then on her movements were inferred from the messages sent by her radio operator and from the accounts of survivors. The first instalment of misfortune came when or soon after she turned to the west, when a tremendous following sea hit the stern doors of the car deck and buckled them in. The captain's first intention was to return to Loch Ryan, going stern first and using the bow-rudder with which the ship was fitted. But it was found that the rudder could not be operated since the pin that held it in position could not be removed, so this avenue of escape was not open; this was the second instalment of misfortune.

Two hours after leaving Stranraer a radio message was received on shore to say that the ship was again off the mouth of Loch Ryan. 'Vessel not under command. Urgent assistance of tugs required.' But there were no tugs to be had; the nearest ones were sheltering from the storm in Douglas Bay off the Isle of Man. The ship had therefore to manage on her own. By now she was listing 10 degrees, partly because the cargo on the car deck had shifted towards the starboard side and the deck itself was under five feet of water. By degrees this seeped through into the passenger and lounge accommodation. At about 10.30 am the likelihood that the vessel would not remain afloat was accepted; the passengers were given life jackets and the boats were prepared for lowering.

Meanwhile the Portpatrick lifeboat had been launched, but failed to find the *Princess Victoria*, which had been broadcasting its position inaccurately, which in the desperate circumstances was not surprising. The ship was drifting to the south-west, so the lifeboat headed in the wrong direction. Soon after 1 pm another radio message reported that the engines were now stopped. Meanwhile the Donaghadee lifeboat and a naval destroyer from Rothesay, far to the north, were on their way to the supposed position of the ship. Just before 2 pm another message was sent to say that she was about six miles east of the entrance to Belfast Lough. Several ships were sheltering there, and one set out to find the stricken vessel, others following some time later. One of the latter, the *Orchy*, ran into floating wreckage where a single lifeboat had managed to survive. It was not possible to rescue its 42 occupants, all male passengers, there and then, so the ship manoeuvred until the boat was under its lee; eventually the lifeboats arrived and took the survivors to safety. All 49 officers and men of the crew and 85 passengers had perished.

At the subsequent Inquiry, held at Belfast seven weeks later, the questions of cause and responsibility were considered. In June 1954 the assessor's Report was issued. The ship was held to have foundered because the stern doors had not been strong enough to withstand the force of the seas and because, after the sea had come through on to the car deck, there had been no provision for clearing it away, so that the ship eventually capsized and sank. Ultimate responsibility was fastened on the British Transport Commission, and specifically upon three successive Marine Superintendents who had had the management of the ship since its delivery to the Scottish Region of the Commission. There was an appeal against the findings, as a result of which the first of the indicted men had the charges against him withdrawn; the second was held to be primarily responsible, in that he did not 'appear to have realised the full implications of the exacting standard of skill and care which his office and its responsibilities demanded of him respecting the seaworthiness of *Princess Victoria*.'

Medals and distinctions were duly awarded to the chief persons who had taken part in the rescue work, and David Broadfoot, the radio officer of the *Princess Victoria*, who had stayed at his post to the last in order to transmit messages, was given the George Medal posthumously.

Dromore (Gatehouse) station as viewed in 1989, now a private dwelling house.
Julia Cheetham

Index